RAND NATIONAL DEFENSE RESEARCH INSTITUTE

T0306306

EXPEDITIONARY CIVILIANS

Creating a Viable Practice of Department of Defense Civilian Deployment

Molly Dunigan, Susan S. Everingham, Todd Nichols, Michael Schwille, Susanne Sondergaard

Prepared for the Office of the Secretary of Defense

For more information on this publication, visit www.rand.org/t/RR975

Library of Congress Cataloging-in-Publication Data
ISBN: 978-0-8330-9199-4

Published by the RAND Corporation, Santa Monica, Calif.

Cover design by Tanya Maiboroda.

Preface

U.S. Department of Defense (DoD) Directive 1404.10 (2009) mandates reliance on military and civilian capabilities to meet national security requirements and requires identification of a subset of civilians to be organized, trained, and equipped to respond to expeditionary requirements. This policy is based on a statutory requirement set forth in Section 854 of the National Defense Authorization Act for Fiscal Year 2007 to identify a deployable cadre to support contingency program management. The policy was reiterated in the 2010 *Quadrennial Defense Review Report*, which named deployable civilians as an enabling capability and mandated the program's improvement and expansion, and the 2012–2013 Strategic Management Plan, which referred to "increased readiness" of deployable civilians as a business goal.

DoD policy on expeditionary civilians has yet to be fully implemented, however. This report presents the results of an end-to-end review and analysis of DoD civilian deployment intended to inform DoD's practices and processes for utilizing civilians to meet the needs of the U.S. combatant commands. The research assesses the viability of current DoD civilian deployment practice in meeting the goals of existing civilian deployment policy and provides a systematic approach to developing and maintaining DoD's civilian deployment capability to ensure that this capability is available to meet combatant commander requirements as efficiently and effectively as possible. To achieve these aims, RAND researchers interviewed more than 80 officials from DoD, the military services, the combatant commands, and analogous U.S. and foreign government organizations that deploy civilians. Using the data obtained from these interviews, along with a policy and literature review, RAND analyzed gaps between policy and practice, reviewed combatant command requirements for expeditionary civilian capabilities, and identified best practices and lessons learned from analogous organizations' civilian deployment practices that could be used to inform the future practice of DoD civilian deployment.

This research was conducted in 2014, and the findings were current as of mid-2015. This report should be of interest to policymakers and military officials both in the United States and abroad who are interested in the issue of civilian deployment, as well as researchers and policymakers interested in workforce mix issues.

This research was sponsored by the Office of the Deputy Assistant Secretary of Defense for Civilian Personnel Policy and conducted within the Forces and Resources Policy Center of the RAND National Defense Research Institute, a federally funded research and development center sponsored by the Office of the Secretary of Defense, the Joint Staff, the Unified Combatant Commands, the Department of the Navy, the Marine Corps, the defense agencies, and the defense Intelligence Community. For more information on the RAND Forces and Resources Policy Center, see www.rand.org/nsrd/ndri/centers/frp or contact the director (contact information is provided on the web page).

Contents

Preface iii
Figures vii
Tables ix
Summary xi
Acknowledgments xxix
Abbreviations xxxi

CHAPTER ONE
Introduction 1
Background 1
Objectives of This Study 8
Study Approach 11
Organization of This Report 14

CHAPTER TWO
Civilian Deployment in Policy Versus Practice 15
DoD Policy on "Expeditionary Civilians" 16
Total Force Management 21
Strategic Human Capital Planning Policy 40
Ability to Meet Operational Needs 50

CHAPTER THREE
Combatant Command Utilization of DoD Expeditionary Civilians 53
Defining the Steps in the Process 54
Combatant Commands' Varied Utilization of Expeditionary Civilians 71

CHAPTER FOUR
Alternative Civilian Deployment Models 79
Selecting Cases of Civilian Deployment Analogous to the DoD Experience 80
Overview of Case Characteristics 82
Typology of Deployment Models 96
Lessons from the Four Deployment Models 107

CHAPTER FIVE
Conclusion and Recommendations: A Framework for the Future 111
Conclusions 111
Recommendations 112

APPENDIXES
A. Glossary of Key Terms 129
B. Organizational Affiliations of Interviewees 131

Bibliography 133

Figures

S.1. Tiered Sourcing Model xxv
2.1. The Total Force Concept as a Set of Overlapping, Scalable Workforces Offering Distinct Capabilities 24
2.2. Center for Strategic and Budgetary Assessment Estimates of Costs per Service Member Deployed 36
3.1. Global Force Management Allocation Process 57
3.2. Joint Individual Augmentation Process Flow Diagram 59
3.3. USCENTCOM Area of Responsibility Relative to Those of Other Geographic Combatant Commands 72
4.1. Reactive Internal Sourcing Model 99
4.2. Proactive Internal Sourcing Model 101
4.3. Reactive External Sourcing Model 103
4.4. Proactive External Sourcing Model 105
5.1. Ownership of Relevant Aspects of Civilian Deployment 114
5.2. The Benefits of Scenario-Based, Mission-Specific Forecasting 119
5.3. Tiered Sourcing Model 127

Tables

S.1. Deployment Models, by Agency xxii
1.1. Organizations Interviewed for This Study 12
2.1. Prior U.S. Air Force Calculations of Relative Costs of Military Versus Civilian Personnel at Home and for a Six-Month Deployment to OIF/OEF 38
2.2. Top 30 Deployable Civilian Occupational Series 49
4.1. Organizations Interviewed for This Study 81
4.2. Characteristics of Organizations Interviewed 87
4.3. Deployment Models, by Agency 97

Summary

The use of U.S. government civilians in overseas contingency operations underwent significant changes during Operation Iraqi Freedom and Operation Enduring Freedom in Afghanistan, particularly during President George W. Bush's second term in office. During his first term, relatively small numbers of civilians from the U.S. Department of State (DoS), the U.S. Agency for International Development (USAID), and other agencies deployed to these contingencies. They worked for such DoD organizations as the Coalition Provisional Authority in Iraq or with Provincial Reconstruction Teams (PRTs) in Iraq and, to a lesser extent, Afghanistan. During this time, however, few *DoD* civilians deployed to support the wars.

In President Bush's second term, it became clear that the United States was not going to make a quick exit from either theater; DoD prioritized reconstruction and stabilization efforts as core missions, DoS asked DoD for help staffing PRTs in Iraq, and the surge intensified the operational tempo that had already exceeded target utilization rates of the active and reserve components of the uniformed military.[1]

After examining options to relieve pressure on the uniformed military, civilian leadership in the Office of the Secretary of Defense (OSD) identified DoD's approximately 700,000-person civilian workforce as a

[1] Note that while overall demand for deployed civilians was driven by manpower needs in the concurrent conflicts in Iraq and Afghanistan during this period, the primary driver of demand for civilians to staff PRTs was the Iraq War, in which PRTs were led by DoS and staffed with both civilians and military personnel. PRTs in Afghanistan during this period were staffed primarily by military personnel.

viable source of deployable personnel, and the first cohort of DoD civilians deployed to Iraq in the summer of 2007. Subsequently, three other forces increased the demand for DoD expeditionary civilians:

- a DoD insourcing initiative that began in 2008 as a result of growing concern over the use of private contractors in war zones
- the need for personnel to assist the Iraqi ministries in 2008
- the increased use of "nonstandard" sourcing solutions within the Global Force Management process to fill U.S. Central Command (USCENTCOM) requests for capabilities not being met by available "assigned" forces or forces already "allocated."

However, the increased demand for civilians was not matched by increased supply. There were few career-based incentives for civilians to agree to deploy or for their employers to let them go. In addition, each service was responsible for recruiting and training civilians from within its ranks, and some efforts were half-hearted.

Collectively, these increases in demand and curtailments in supply forced a transformation. In January 2009, DoD Directive (DoDD) 1404.10 established the Civilian Expeditionary Workforce (CEW), envisioned as a standing cadre of 20,000–30,000 civilians already organized, trained, cleared, and equipped to mobilize quickly as needed. The directive partially transferred DoD civilian sourcing responsibility from the services to the Office of the Deputy Assistant Secretary of Defense for Civilian Personnel Policy (CPP).[2]

Over the next five years, the CEW Program Office (CEWPO) in CPP would recruit, train, and deploy nearly 5,000 civilians to a variety of theaters. But its dual role (both policy and operations, particularly in the form of sourcing) proved challenging, as did the attempt to transform the way DoD managed its civilian workforce. With no new resources to help manage its new responsibilities, CPP ultimately ceased its role in sourcing expeditionary civilians and began to review the plan for a large standing cadre.

---

[2] U.S. Department of Defense Directive 1404.10, *DoD Civilian Expeditionary Workforce*, January 23, 2009.

Looking toward the future, the problem that the deployment of expeditionary civilians is meant to solve is that the U.S. government currently lacks deployable capacity to support complex contingency operations with the skill sets in which DoD civilians are qualified, and in which they may even enjoy a comparative advantage over other types of personnel. This is particularly true for civilian requirements not already assigned to a force, such as joint requirements labeled as "civilian acceptable." This lack of capacity is exacerbated by ongoing efforts to downsize the U.S. military and insource functions formerly reserved for contingency-support contractors. However, the nature, scale, and location of potential future contingency operations are currently unknown. In the midst of such uncertainty, the CEW construct was developed to provide a scalable source of such individuals.[3]

In March 2014, the Army G-1 (Manpower and Personnel) began fulfilling all CEW operational responsibilities. The Army G-1 tried to close gaps in the CEW sourcing process and reduce delays in deployment; however, this model also had shortcomings. Shortcomings in the CEW sourcing process led USCENTCOM to create its own CEW officc in mid-2014 to handle the operational side of civilian deployment for the USCENTCOM area of responsibility (AOR)—the primary combatant command (CCMD) customer of expeditionary civilian capabilities to date.

Study Objectives

We were asked to assess CPP's decisions to (1) focus solely on its policy role with regard to expeditionary civilians; (2) have the military services retain operational responsibility for DoD expeditionary civilians; and (3) shift away from the large-scale CEW "cadre" model envi-

---

[3] A notable constraint on this construct is that the primary role of DoD civilians capable of serving in an expeditionary role is not to prepare for or support contingency operations but to execute the DoD components' day-to-day responsibilities under Title 10 of the U.S. Code (U.S.C.). This constraint translates to combinations of incentives—for both the expeditionary civilians themselves and their home offices—that do not necessarily motivate seamless civilian deployment processes, as discussed in further detail throughout this report.

sioned in DoDD 1404.10 (2009) to a more reactive model designed to respond to requirements for expeditionary civilians as they arise. Based on this request, the objectives of this study were twofold: to conduct an end-to-end review of guidance over the process of DoD civilian deployment and to recommend guidelines for establishing and maintaining a civilian deployment capability that meets the requirements of the CCMDs for various regions. In pursuing these objectives, we took a *prospective, forward-looking* view of the possibilities for a civilian deployment framework that will continue to be viable in ten to 20 years, rather than a retrospective view examining the issues surrounding the implementation of the CEW and associated policy thus far.

Study Approach

We worked with DoD officials to identify goals and potential gaps in current policies and planning. To do so, we first performed an end-to-end review of existing policies and procedures relevant to DoD civilian deployment, existing policy analysis and relevant reports by such organizations as the U.S. Government Accountability Office (GAO), and secondary literature on civilian deployment. We then interviewed more than 80 officials in 45 offices across DoD, the Joint Staff, the CCMDs, the U.S. military, other U.S. government organizations, and foreign governments. Using this information, we

- assessed the viability of the current civilian deployment process in meeting the goals outlined in policy
- reviewed the CCMDs' requirements for and perspectives on expeditionary civilians to establish baseline manpower and personnel management requirements
- compared the civilian deployment policies and practices of both other U.S. government agencies and foreign government organizations to glean best practices and lessons learned
- made recommendations for policy, planning, and operational actions that could be taken to help ensure a long-term civilian deployment capability.

It is critical to note at the outset that over the course of this research, we encountered some difference of opinion among our interviewees regarding whether DoD civilians should be deployable at all—and if so, under what circumstances. Some interviewees argued that civilians are not as mentally prepared for deployment as military personnel, that military or DoS personnel can fill most positions (and the military should be grown as needed), and that protecting civilians in hostile environments is an additional burden on the military. In addition, existing policies are somewhat unclear on the issue of civilian deployment. In response to guidance from our research sponsor, this report does not address the question of suitability of civilians for deployment and instead proceeds from the assumption that DoD aims to continue deploying civilians and believes that civilians are deployable (at least in certain situations). However, the fact that such debate exists throughout the defense community is notable, and DoD should consider this larger question in future deliberations regarding the direction of the civilian deployment concept.

This research was conducted in 2014, and the findings were current as of mid-2015.

Policy and Gap Analysis

For the end-to-end review of policies and procedures associated with DoD's civilian deployment capability, we performed a gap analysis of the discrepancies between policy and practice using policy documents and interviews to examine existing policies on civilian deployment, what those policies are missing, and how gaps between policy and practice may inform the design of a workable civilian deployment model. The interviews, in particular, highlighted the following as critical considerations:

- the larger context of total force management
- cost-efficiency
- recruitment, retention, and development of the DoD civilian workforce
- the ability to meet operational needs.

In exploring these critical issues, it became clear that three bodies of policy are relevant to the topic of civilian deployment, so we assessed the literature and policy guidance on both Total Force management and strategic human capital planning in addition to existing DoD policy on civilian deployment.

DoD Policy on Civilian Deployment

DoDD 1404.10, which created the CEW in 2009, was the reigning policy on civilian deployment when this research was conducted. It stated that it is DoD policy to rely on a mix of capable military members and DoD civilian employees to meet global national security mission requirements. Civilians trained in certain job specialties and positions, such as contracting, security, public affairs, law, information technology, human resources, engineering, and translation, would support DoD combat contingencies, emergency operations, humanitarian and civic assistance activities, disaster relief, restoration of order, drug interdiction, and stability operations.

Several structural changes pertaining to DoD civilian workforce management ensued, but that vision never fully came to fruition for the reasons discussed earlier. As a result, the International/Expeditionary Support (IES) Office (formerly the CEWPO) within CPP was working to revise DoDD 1404.10 at the time of this writing. Meanwhile, the large-scale standing cadre of preidentified employees has been forgone in favor of a smaller force totaling approximately 5,300 across DoD.

Discrepancies Between Policy and Practice

The initial impetus for the CEW was to fill low-density, high-demand positions not easily filled by the uniformed military. But in practice, expeditionary civilians were used to reduce stress on the uniformed military, reduce the military's reliance on contractors, and circumvent force management levels that precluded the deployment of additional uniformed military personnel to a given theater. While force management levels presented a quantifiable and easily identified challenge that expeditionary civilians were used to surmount, the overarching narrative here is that the U.S. government simply did not have sufficient deployable capacity to meet its needs during this period. Since 2009, it appears that the bulk of DoD expeditionary civilians have performed jobs that could have been done by the military, had sufficient troop capacity existed. This fact speaks to the need to fully comprehend civilian deployment in the context of the "Total Force."

Total Force Management

The notion of the Total Force originated in policy on August 21, 1970, with a memorandum from Secretary of Defense Melvin Laird. The original intent was quite broad and included allied and host-nation forces in occupied areas, as well as U.S. forces. However, the Total Force policy eventually came to be associated almost exclusively with the integration of the reserve components into the active military services. More recent DoD policy—elaborated, for instance, in the *Quadrennial Defense Review Report* beginning in 2006—defines the Total Force as consisting of military, civilian, and contractor personnel.

Three issues related to management of the Total Force are relevant to both the motivation for and the long-term feasibility of DoD's expeditionary civilian capability:

- *Civilian ability to work across the joint workforce.* DoD civilians are currently "owned" by a particular military service or DoD component (Army, Navy, Air Force, Fourth Estate). They must apply to deploy with another service, and the switch can hinder their career progression. The service for which they work must allow their departure, which it is typically reluctant to do because of the lack of backfill available for positions vacated by civilian deployees, as well as the fact that the home office originally employing the individual continues to pay the individual's salary while he or she is deployed.
- *Insourcing of numerous capabilities previously outsourced to contractors.* In the mid-2000s, several high-profile incidents regarding contractor transparency, accountability, and overbilling in Iraq and Afghanistan called extensive outsourcing into question. In 2008, Deputy Secretary of Defense Gordon England directed the use of DoD civilians for new functions rather than contractors when the job duties and cost efficiencies made it sensible to do so.

- *The cost-efficiency of deploying civilians as opposed to military personnel or contractors for various tasks.* Interestingly, our interviews uncovered a puzzling discrepancy in DoD and other officials' understanding of the comparative costs of deploying military personnel versus DoD civilians: Several interviewees argued that military personnel are more expensive to deploy, while others were adamant that civilian personnel are more costly to deploy. Our subsequent literature review on the topic uncovered little existing formal analysis comparing the cost to deploy a civilian with the cost to deploy military personnel. Better understanding of these relative costs across the department would assist policymakers and other DoD officials in understanding the best circumstances under which to deploy civilians versus military personnel.

Strategic Human Capital Planning Policy

A third body of policy relevant to civilian deployment—and integrally related to Total Force policy—is strategic human capital planning policy.[4] Each service must periodically map the array of competencies and performance criteria it requires and improve processes to achieve those standards, as discussed in detail in Chapter Two of this report.

Adequate planning is necessary to ensure that DoD can draw on appropriate numbers of expeditionary civilians with the correct skill sets for various contingencies on short notice if needed, either from the existing DoD workforce or from preidentified outside sources.

Relatedly, it is essential that incentives—both to encourage the civilian to deploy and to encourage the home office to support that deployment—be correctly aligned to encourage expeditionary civilian deployment. One problem highlighted over the course of this research is that home offices do not receive support to backfill positions left vacant by a deploying civilian. Because of this, the service or component is motivated to deter the employee from deploying and to be

---

[4] Strategic human capital planning is an essential component of Total Force management and involves strategic management of DoD's overall human capital, including the active and reserve components, DoD civilians, and the contractor workforce.

reluctant, in general, to offer up its civilians for expeditionary positions. Meanwhile, the other components are powerless to force a particular service to volunteer its civilian personnel for deployment. We therefore identified a need for joint oversight by a nonservice entity empowered with the capability to pull civilians from across the services and components into expeditionary positions, particularly if DoD wants to ensure the continued existence of a viable, robust expeditionary civilian capability.

Deployable Civilian Demand Signal

Specific human capital planning guidance pertaining to expeditionary civilians is included in 10 U.S.C. 2333, "Joint Policies on Requirements Definition, Contingency Program Management, and Contingency Contracting," (C)(3) and (f)(5), which requires "identification of a deployable cadre of experts, with the appropriate tools and authority, and trained in processes under paragraph (6)." However, GAO noted in 2012 that this requirement had not been implemented. In response to this report, the IES Office developed a "demand signal" for expeditionary civilians based on CCMD requirements. The demand signal is "the number of civilians in particular career fields who should be available to deploy in support of a broad range of expeditionary requirements." At the time of this writing, the Defense Civilian Personnel Advisory Service and personnel from the IES Office were in the process of launching the demand signal.

Combatant Command Utilization of DoD Expeditionary Civilians

The operational side of DoD civilian deployment comprises several elements: recruiting, screening, and selecting candidates for deployable positions (sourcing), predeployment training and obtaining medical and security clearances (readiness preparation), and tracking civilians during and after deployment to keep them apprised of related announcements, health screenings, or awards or benefits.

Little detailed analysis has been done on the CCMDs' need for and understanding of expeditionary civilian capabilities. To fill this gap, we analyzed (1) the operational process through which expeditionary civilians are tasked to the CCMDs, are prepared to deploy, and are tracked during and after deployment; (2) CCMD requirements for expeditionary civilians in recent years; and (3) CCMD perspectives regarding the utilization of expeditionary civilians.

In short, we found that the CCMDs vary widely in their projected use of expeditionary civilians and their understanding of expeditionary civilian capabilities, and that sourcing delays are a key reason CCMDs do not view expeditionary civilians as a reliable labor source. These findings highlight lessons for the future operation of DoD civilian deployment. See Chapter Three for a complete discussion of the different operational models for civilian deployment that DoD has employed in recent years.

CCMD Requirements for and Perspectives on Utilization of Expeditionary Civilians

Since the creation of the CEW in 2009, USCENTCOM has by far been the largest CCMD customer of expeditionary civilian capabilities, generating 89 percent of requirements for expeditionary civilians as of mid-2014. This is unsurprising, given the two concurrent wars in Iraq and Afghanistan occurring within the USCENTCOM AOR during this time period. As requirements to deploy civilian personnel continued to rise during this period, the recruitment and deployment process became untenable by the Army, and, as a result, all of the services were required to provide capabilities to USCENTCOM to meet the increased demand.

In recent years, a large degree of uncertainty still exists regarding future requirements for expeditionary civilians in different locations and across mission sets beyond those in the USCENTCOM AOR. This is particularly so because the USCENTCOM AOR represents a relatively small region in comparison to those represented by the other geographic CCMDs. However, continuing conflict in the region may perpetuate or generate a need for expeditionary civilians. The most pressing question is whether unforeseen demands will arise in other areas of the globe in addition to those in the USCENTCOM AOR.

Our interview findings indicate that the CCMDs vary quite drastically in their actual utilization of, and overall perspectives regarding, expeditionary civilians. We found that the CCMDs tend to fall into one of three categories in this regard. First, several CCMDs argue that they have absolutely no need for expeditionary civilians. Second, some CCMDs argue that utilization of expeditionary civilians is either administratively burdensome or takes too long to be operationally useful, suggesting they could use civilians but are reluctant to do so. Third, only one CCMD (USCENTCOM) uses DoD expeditionary civilians fairly extensively, subscribing to the practice of utilizing them simply as additional "boots on the ground."

To develop a viable civilian deployment concept over the long term, it is therefore advisable that DoD consider the potential needs of other CCMDs beyond USCENTCOM for expeditionary civilian capabilities and, relatedly, how to broaden CCMD utilization of these capabilities.

Alternative Civilian Deployment Models

To find the most useful lessons from the civilian deployment models used by analogous organizations, we cast a wide net, interviewing representatives of several DoD agencies that have their own well-established civilian deployment programs, as well as officials from the U.S. Department of State, U.S. Department of Homeland Security, and USAID, along with representatives from defense organizations in the United Kingdom, Canada, Australia, and the European Union.

Overall, we identified four models that these 17 analogous organizations applied to deploy civilians. The models differ along two main dimensions: the extent to which they source individuals to deploy from within the organization's existing civilian workforce (internal sourcing), as opposed to searching for candidates external to the organization (external sourcing), and the extent to which the organizations establish a pool of preidentified individuals prior to the issuance of requirements (proactive sourcing), as opposed to identifying candidates for positions after requirements are issued (reactive sourcing). In Table S.1, we categorize the 17 analogous organizations into four deployment models.

The four models highlight the differences in how organizations handle civilian deployment. We found it notable that among the agencies we examined, those that applied a proactive sourcing approach were able to deploy more quickly than those that began the recruitment process upon request. Meanwhile, organizations that relied on external sourcing models encountered numerous rules and regulations. Furthermore, regardless of whether individuals were sourced internally or externally, some type of oversight organization was necessary to ensure successful deployments. We also identified opportunities for

Table S.1
Deployment Models, by Agency

	Reactive Sourcing	Proactive Sourcing
Internal Sourcing	U.S. Customs and Border Protection Drug Enforcement Administration Defense Intelligence Agency Canadian Department of National Defence DoS Bureau of South and Central Asian Affairs U.S. Department of Homeland Security Office of International Affairs	Defense Contract Management Agency Defense Logistics Agency Federal Emergency Management Agency DoS Bureau of Diplomatic Security UK Ministry of Defence, Support to Operations Australian Defence Science and Technology Organisation (now the Defence Science and Technology Group)
External Sourcing	Afghanistan and Pakistan Strategic Partnership Office European External Action Service, Civilian Planning and Conduct Capability	USAID Office of Transition Initiatives USAID Crisis Surge Support Staff DoS Bureau of Conflict and Stabilization Operations European External Action Service, Civilian Planning and Conduct Capability, Civilian Response Teams

distinct organizations to share enabling capabilities for civilian deployments, such as predeployment training and medical facilities.

Summary of Findings and Recommendations

Overall, we found that the current DoD civilian deployment concept requires certain modifications to be viable over the long term. We reached this conclusion for several reasons. First, aspects of the sourcing model in the current deployment process—which envisages decentralized recruitment, screening, and selection of civilian candidates by the military services for deployable position requirements—did not work in the period prior to the drafting of DoDD 1404.10, largely because the services were not adequately motivated to encourage or allow civilian deployment without backfill support for positions left vacant by deployees. In the absence of some higher-level organization with visibility across all DoD components—perhaps at the OSD or Joint Staff level—to push the services to offer up their civilians for these tasks, supply of expeditionary civilian capabilities remains unlikely to be adequate to meet demand. This is especially true because the individual services lack the power to pull civilians from other services. This, in turn, provides little incentive for civilian personnel to volunteer for various expeditionary positions, because their careers—at least in their current positions—are unlikely to benefit from their voluntary deployment.

Second, USCENTCOM's sourcing model, which entails having a separate CEW office housed at the CCMD responsible for the recruitment, screening, and selection of civilian candidates, is not entirely applicable to other CCMDs. Because we were asked to develop a plan for the broad-based use of expeditionary civilian capabilities across the CCMDs for the next several decades, this inapplicability of the USCENTCOM model to the other CCMDs poses a concern.

Third, the extent to which military plans actually consider expeditionary civilian requirements is questionable at this point, at least as indicated by our interviews.

Fourth, many CCMDs appear unlikely to utilize expeditionary civilian capabilities in the future, as they are unaware of the benefits that expeditionary civilians can provide, are wary of potential sourcing delays that could hinder any operational plans relying on expeditionary civilians, or feel that they have no need for additional expeditionary civilian capabilities beyond their own current capacity.

Finally, as noted earlier, our interviews uncovered conflicting perceptions across DoD regarding the respective costs of deploying DoD civilians versus U.S. military personnel.

Our research indicates that these issues could be remedied through the adoption and effective implementation of the following recommendations.

Joint oversight at the Joint Staff or OSD level, or by a specially designated executive agent. We identified several key operational responsibilities as being suited to a joint-level organization of some sort: assignment of requirements, sourcing, tracking, and readiness preparation of DoD civilians not deploying to support service missions. While we make no claim as to the best joint-level "owner" of these functions, we outline the pros and cons of several potential scenarios in Chapter Five of this report.

The pursuit of mission-specific, scenario-driven forecasting to inform manpower planning. We strongly recommend more extensive manpower planning and forecasting pertaining to future requirements for DoD expeditionary civilians. Such planning is needed to (1) delineate appropriate numbers of Emergency-Essential (E-E)–coded personnel and enact E-E coding to a greater extent, and (2) to help reduce service uncertainty regarding the number of service civilians who may have to deploy in any given contingency.[5] The Defense Civilian Personnel Advisory Service's "demand signal" work (elaborated in Chapter Two), is a strong step in the right direction in this regard, but more remains to be done.

[5] E-E personnel constitute one of the three main categories of expeditionary civilian personnel outlined in DoDD 1404.10, 2009. E-E personnel are defined as "a position-based designation to support the success of combat operations or the availability of combat-essential systems."

Creation of a tiered sourcing model combining internal and external, proactive and reactive hiring mechanisms to ensure readiness and flexibility in the force across a spectrum of possible future contingencies. The comparative case analysis and interviews conducted for this study indicate that CCMD utilization of expeditionary civilian capabilities will occur only if the capability is viewed as reliable, and expeditionary civilian capabilities will be viewed as reliable only if individuals are able to deploy relatively quickly. Moreover, our comparative analysis of analogous cases indicates that expeditionary civilians will likely be ready to deploy on short notice only if they are proactively sourced *to at least some extent.* This does *not* necessarily have to be a large-scale cadre as originally envisioned for the CEW, however. It could be a "tiered approach" involving different categories of personnel. A tiered sourcing approach is suited to organizations that have a need to deploy a wide range of civilians within differentiated time scales.

Figure S.1 depicts a tiered sourcing model. At the top is the readiness pool, which consists of personnel ready to deploy on short notice. Personnel in this pool could be sourced internally or externally and

Figure S.1
Tiered Sourcing Model

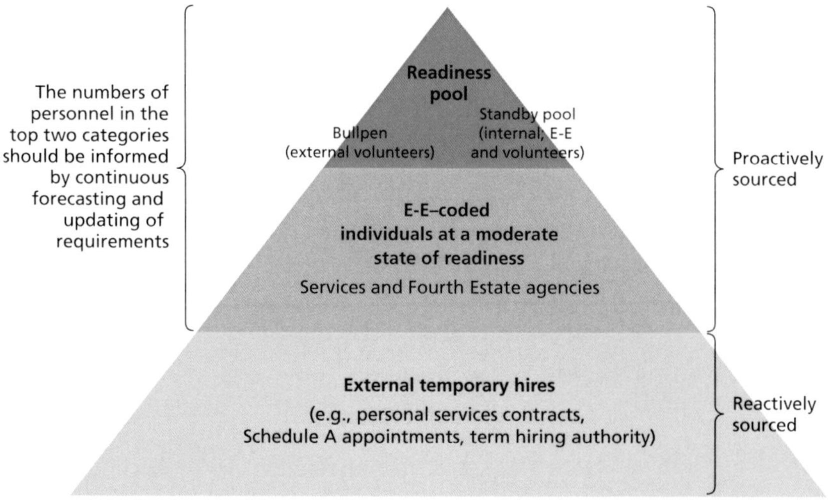

RAND *RR975-S.1*

could be either E-E–coded or volunteer full-time employees, or temporary employees sourced externally. The second tier of the model comprises E-E–coded individuals. Personnel at this level will be in positions coded to deploy if a requirement arises that fits their particular skill sets. These personnel, however, will not be kept at the same level of readiness as those in the readiness pool. The last tier of the model consists of external temporary hires. These individuals will be hired from outside the organization and will be used to fill requirements only once the top two tiers have been fully utilized. These individuals will be term employees hired only when surge requirements arise, outstripping capacity within the top two tiers.

Raising the awareness of expeditionary civilian capabilities in CCMDs. Raising awareness serves two primary functions. First, it provides those unfamiliar with civilian deployment with information about what it means to be a deployed civilian. This gives potential applicants a realistic preview of the roles in which they could possibly deploy and helps adjust expectations for both those generating requirements and the civilians who consider volunteering for a deployment. Second, raising awareness ensures that individuals making requests for deployed civilians understand both the current and future capabilities available. This helps those individuals generating requirements understand where the utilization of civilian deployees would be most appropriate and how they could be employed. Such initiatives thus support a cultural shift in overcoming potential misconceptions about the availability, deployment process, and usability of civilians in support of military operations.

A comparative cost study clarifying the relative costs of deploying civilians, military personnel, and contractors for particular tasks, to inform Total Force Management policy, inform force sizing initiatives, and ensure a common understanding of which force is most cost efficient in a given scenario. As noted previously, there is a dearth of analysis comparing the relative costs of *deploying* civilians, contractors, and military personnel. Because such information is critical to the development of a cost-efficient Total Force policy specifying distinct roles for civilians, contractors, and military personnel deployed to a theater of conflict, we recommend that the policy owner of expeditionary civilian capabilities conduct or oversee a comparative cost study along these lines. The results of this comparative cost study could then be used to inform force sizing initiatives across the services, CCMDs, and Joint Staff. We suggest a three-step method for such a study to capture numerous variables simultaneously in a robust manner:

- Develop a comprehensive cost structure considering the security costs for various personnel types, as well as salaries and associated benefits.
- Create a data set of representative military, civilian, and contractor personnel characteristics to control for experience and other skills/characteristics simultaneously with cost (to know where more expensive personnel are actually more effective and thus worth the price).
- Consider scenarios in which each group would be most cost-effective.

Acknowledgments

We gratefully acknowledge the assistance of a number of individuals across the U.S. Department of Defense, Joint Staff, combatant commands, U.S. military services, other U.S. government agencies, and foreign governments who took the time to speak with us for the purposes of this study. Although we cannot name them publicly, we are extremely grateful for their assistance. At RAND, we thank Laura Novacic for her administrative support in preparing this document, Melissa Bauman for her writing support, Dick Hoffmann for his support in identifying interviewees at the combatant commands, and Samantha Bennett for her editorial support. We also thank John Winkler, Lisa Harrington, and Jennifer Lamping-Lewis for their management support over the course of the study, and Ginger Groeber, Wade Markel, and Agnes Schaefer for their careful review of this report. The Office of the Deputy Assistant Secretary of Defense for Civilian Personnel Policy deserves special thanks for funding this research, and we would especially like to thank the Deputy Assistant Secretary of Defense for Civilian Personnel Policy, Paige Hinkle-Bowles, as well as Julie Blanks and Joe Daniel, for the guidance they provided.

Abbreviations

AOR	area of responsibility
APSPO	Afghanistan and Pakistan Strategic Partnership Office
CBP	U.S. Customs and Border Protection
CEW	Civilian Expeditionary Workforce
CEWPO	Civilian Expeditionary Workforce Program Office
CCMD	combatant command
COM	chief of mission
CONUS	continental United States
CPCC	Civilian Planning and Conduct Capability
CPP	Civilian Personnel Policy
CS3	Crisis Surge Support Staff
CSDP	Common Security and Defence Policy
DCMA	Defense Contract Management Agency
DCPDS	Defense Civilian Personnel Data System
DEA	Drug Enforcement Administration
DIA	Defense Intelligence Agency

DLA	Defense Logistics Agency
DoD	U.S. Department of Defense
DoDD	U.S. Department of Defense directive
DoDI	U.S. Department of Defense instruction
DoS	U.S. Department of State
DRRS	Deployment Readiness Reporting System
DSTO	Defence Science and Technology Organisation [now the Defence Science and Technology Group]
E-E	Emergency-Essential
EEAS	European External Action Service
FEMA	Federal Emergency Management Agency
FY	fiscal year
GAO	U.S. Government Accountability Office
GFM	Global Force Management
GFMAP	Global Force Management Allocation Plan
IES	International/Expeditionary Support
IT	information technology
JIA	Joint Individual Augmentation
JMD	joint manning document
JTF	joint task force
MOD	UK Ministry of Defence
NATO	North Atlantic Treaty Organization
NCE	Non-Combat Essential

NDAA	National Defense Authorization Act
OEF	Operation Enduring Freedom
OIF	Operation Iraqi Freedom
OTI	Office of Transition Initiatives
OUSD(P&R)	Office of the Under Secretary of Defense for Personnel and Readiness
PRT	Provincial Reconstruction Team
QDR	Quadrennial Defense Review
RFF	request for forces
S2O	Support to Operations
SHCP	strategic human capital planning
USAFRICOM	U.S. Africa Command
USAID	U.S. Agency for International Development
U.S.C.	U.S. Code
USCENTCOM	U.S. Central Command
USEUCOM	U.S. European Command
USSOCOM	U.S. Special Operations Command
USSOUTHCOM	U.S. Southern Command
USSTRATCOM	U.S. Strategic Command
USTRANSCOM	U.S. Transportation Command

CHAPTER ONE

Introduction

Background

The use of U.S. government civilians in overseas contingency operations changed significantly during Operation Iraqi Freedom (OIF) and Operation Enduring Freedom (OEF) in Afghanistan, particularly during President George W. Bush's second term in office. During President Bush's first term, U.S. government civilians from various agencies, including the U.S. Department of State (DoS) and the U.S. Agency for International Development (USAID),[1] deployed in relatively small numbers to work for U.S. Department of Defense (DoD) organizations in Iraq, such as the Office for Reconstruction and Humanitarian Assistance and the Coalition Provisional Authority, or on Provincial Reconstruction Teams (PRTs) that had been established in Afghanistan in November 2002.[2] During this time, however, few DoD civilians were deployed to support the wars.[3]

---

1 Other agencies participating included the U.S. Department of Agriculture, U.S. Department of Justice, U.S. Department of Health and Human Services, U.S. Department of Commerce, and U.S. Department of the Treasury.

2 Note that only three U.S. government agencies were identified as having a civilian position on U.S. military–led PRTs in Afghanistan at this time: DoS, USAID, and the U.S. Department of Agriculture. See Robert Perito, *The U.S. Experience with Provincial Reconstruction Teams in Afghanistan: Lessons Identified*, Special Report 152, Washington, D.C.: United States Institute of Peace, October 2005.

3 The few DoD civilians who did deploy during this time worked at USCENTCOM forward headquarters in Qatar.

In President Bush's second term, it became clear that the U.S. military was not going to make a quick exit from either theater due to rapidly deteriorating security conditions and guidance set forth in DoD Directive (DoDD) 3000.05, which stated that stability operations were "a core U.S. military mission that the Department of Defense shall be prepared to conduct."[4] At approximately the same time, President Bush issued National Security Presidential Directive 44,[5] which charged the Secretary of State with leading integrated U.S. government reconstruction and stabilization efforts; DoS stood up the Office of the Coordinator for Reconstruction and Stabilization to enhance civilian capacity for crises involving complex emergencies; and PRTs were stood up in Iraq to combine "economic, military and political people in teams to help local and provincial governments."[6]

Modeled after the Vietnam War–era Civil Operations and Revolutionary Development Support program, PRTs offered a promising step forward toward sustained interagency cooperation.[7] By the end of 2006, there were seven U.S. PRTs in Iraq, and in January 2007 President Bush announced the creation of an additional ten as part of his "new way forward," otherwise known as the "surge."[8] Moreover, unlike the military-led PRTs in Afghanistan, PRTs in Iraq were to be led by DoS and staffed with civilians to a greater extent. With the expansion of the PRT concept, however, civilian agencies were unable to provide sufficient staffing. Even before the addition of the ten new teams, most PRTs lacked their full complement of civilians because DoS experi-

---

[4] U.S. Department of Defense Directive 3000.05, *Military Support for Stability, Security, Transition, and Reconstruction (SSTR) Operations*, November 28, 2005, p. 2. This directive was updated and formalized as a DoD instruction (DoDI) in 2009.

[5] National Security Presidential Directive 44, *Management of Interagency Efforts Concerning Reconstruction and Stabilization*, December 7, 2005.

[6] Condoleezza Rice, quoted in Cameron Sellers, *Provincial Reconstruction Teams: Improving Effectiveness*, thesis, Monterey, Calif.: Naval Postgraduate School, September 2007, p. 17.

[7] U.S. Joint Forces Command Joint Warfighting Center, "Provincial Reconstruction Teams," Pre-Doctrinal Research White Paper No. 07-01, November 21, 2007, p. 16.

[8] U.S. Government Accountability Office, *Provincial Reconstruction Teams in Afghanistan and Iraq*, Washington, D.C., GAO-09-86R, October 1, 2008, p. 5.

enced "difficulties in recruiting, particularly among senior foreign service officers."[9]

As a result, Secretary of State Condoleezza Rice asked Secretary of Defense Donald Rumsfeld for assistance in staffing 200–250 civilian PRT positions. In response, Rumsfeld directed David Chu, Under Secretary of Defense for Personnel and Readiness, to explore the issue. By February 2007, Patricia Bradshaw, then–Deputy Under Secretary of Defense for Civilian Personnel Policy, had consulted with the Director of the Joint Staff and the Director for Manpower and Personnel in J1 (Manpower and Personnel Directorate) to discuss options for staffing PRT billets.[10]

At first, these discussions explored the possibility of using the reserve component, but the reserves' operational tempo already exceeded policy guidance issued by the new Secretary of Defense, Robert Gates.[11] In fact, a 2007 Defense Science Board report noted that actual dwell times—that is, time between deployments—were 1:1 (one year deployed and one year at home) for the active component and 1:3 (one year deployed and three years at home) for the reserve component and concluded that DoD's policy goal for the utilization rates of the uniformed military was "not achievable without a substantial reduction in deployed tempo."[12] The report added that the policy goal could not be achieved even with the planned increases in Army and

---

[9] Robert Perito, *Provincial Reconstruction Teams in Iraq*, Special Report 185, Washington D.C.: United States Institute of Peace, March 2007; USAID experienced similar recruiting problems.

[10] Interview with researchers conducting a related study at National Defense University's Center for Complex Operations, June 16, 2014.

[11] In a 2007 memo, Secretary Gates confirmed Rumsfeld's policy establishing a 1:2 ratio of deployment time to time at home station for active-component units and a 1:5 ratio of mobilization time to demobilization time for reserve-component units. See Robert Gates, Secretary of Defense, "Utilization of the Total Force," memorandum, January 19, 2007.

[12] Defense Science Board, *Final Report of the Task Force on Deployment of Members of the National Guard and Reserve in the Global War on Terrorism*, Washington, D.C.: U.S. Department of Defense, September 2007, p. 23.

Marine Corps end strength, which sought to add 75,000 active-duty soldiers and 25,000 active-duty Marines by 2013.[13]

These factors led the Office of the Under Secretary of Defense for Personnel and Readiness (OUSD[P&R]) to conclude that the approximately 700,000-person DoD civilian workforce was the most viable option for relieving pressure on the uniformed military. So, a team within Civilian Personnel Policy (CPP) advertised positions, recruited volunteers, and sent 80 resumes of interested volunteers to DoS. Of these, 49 DoD civilians were selected and deployed to Iraq in the summer of 2007.[14]

Although DoD's support to DoS was intended to be a stopgap measure until DoS could provide more people to the PRTs, DoS continued to have difficulty finding volunteers to serve in Iraq. To counter this problem, Condoleezza Rice announced in October 2007 that DoS would order as many as 50 U.S. diplomats to Iraq to serve in the embassy and on PRTs.[15] Had this plan proceeded, it would have been the first large-scale forced assignment since the Vietnam War, but Rice changed her mind as a result of strong internal opposition.[16] Importantly, the DoS hiring problems, although relatively small in terms of numbers needed to staff PRTs, translated into rotational and enduring

---

13 U.S. Government Accountability Office, *Defense Infrastructure: Army and Marine Corps Grow the Force Construction Projects Generally Support the Initiative*, Washington, D.C., GAO-08-375, March 2008.

14 Interviews with researchers conducting a related study at National Defense University's Center for Complex Operations, June 16, 2014. Note that the Office of the Deputy Under Secretary of Defense for Civilian Personnel Policy transitioned to the Office of the Deputy Assistant Secretary of Defense for Civilian Personnel Policy during the period discussed in this report. For simplicity, we refer to this entity as CPP.

15 Karen DeYoung, "State Department to Order Diplomats to Iraq," *Washington Post*, October 27, 2007a.

16 Karen DeYoung, "State Department Won't Order Diplomats to Iraq," *Washington Post*, November 16, 2007c. This opposition manifested in an October 2007 DoS town hall meeting during which hundreds of foreign service officers criticized the forced assignment policy. One officer stated that service in Iraq was "a potential death sentence," and the vice president of the union representing U.S. diplomats declared, "Directed assignments, we fear, can be detrimental to the individual, to the post, and to the Foreign Service as a whole." Quoted in Karen DeYoung, "Envoys Resist Forced Iraq Duty," *Washington Post*, November 1, 2007b.

requirements for DoD civilians. This increased the demand for DoD expeditionary civilians and forced OUSD(P&R) to explore ideas on how to create a more formal structure to recruit, select, and deploy DoD civilians.

The use of DoD civilians in overseas contingency operations was also subject to three other forces which, taken together, greatly increased the demand for DoD expeditionary civilians. The first was the DoD insourcing initiative that began in 2008 due, in large part, to a U.S. Government Accountability Office (GAO) report that stated, "the increased reliance on contractors has raised a number of issues and concerns that warrant continued attention."[17] The report concluded that DoD needed to reexamine its extensive reliance on contractors, particularly for contingency operations in Iraq, so that it would not lose accountability or control over mission-related decisions.[18] The GAO report, and other high-profile contractor malfeasance news stories,[19] prompted Deputy Secretary of Defense Gordon England to issue a memorandum titled "Guidelines and Procedures on In-Sourcing New and Contracted Out Functions." Issued in April 2008, this memo stated, "DoD Components are to ensure consideration is given, on a regular basis, to using DoD civilian employees to perform new functions and functions that are performed by contractors but that could be performed by government employees."[20] The incoming Obama administration shared concerns about the growing use of contractors, and, as

---

[17] David M. Walker, Comptroller General of the United States, *Defense Management: DoD Needs to Reexamine Its Extensive Reliance on Contractors and Continue to Improve Management and Oversight*, testimony before the Subcommittee on Readiness, Committee on Armed Services, U.S. House of Representatives, Washington, D.C.: U.S. Government Accountability Office, GAO-08-572T, March 2008, p. 1.

[18] Walker, 2008, p. 9.

[19] Loren Thompson, "Pentagon Insourcing Binge Begins to Unravel," *Forbes*, March 7, 2011.

[20] Gordon England, Deputy Secretary of Defense, "Implementation of Section 324 of the National Defense Authorization Act for Fiscal Year 2008 (FY 2008 NDAA)—Guidelines and Procedures on In-Sourcing New and Contracted Out Functions," memorandum, April 4, 2008, p. 1.

a result, it made insourcing "a top priority."[21] Although the value and effectiveness of the insourcing initiative is still being debated, one effect insourcing had in 2008–2009 was to increase the demand for DoD civilians in Iraq and Afghanistan. The insourcing initiatives and associated debates are explored in further detail in Chapter Two's discussion of Total Force policy.

The second major force that increased the demand for DoD expeditionary civilians was Lieutenant General Frank Helmick's 2008 request to Patricia Bradshaw to help fill positions to assist Iraqi ministries. Helmick, about to take command of Multi-National Security Transition Command–Iraq, knew Bradshaw well from his time working on Gordon England's staff and was therefore familiar with her role in staffing PRTs. Bradshaw, cognizant of operational tempo stresses on the uniformed military and aware that ministry work required bureaucratic experience, decided to expand the use of DoD civilians to accommodate Helmick's request. So, in the summer of 2008, she took a team to Iraq to visit Multi-National Security Transition Command–Iraq, the Iraqi Ministries, and Multi-National Force–Iraq leadership to determine a list of appropriate "advisory" jobs for civilians. These job descriptions called for capabilities and expertise unrelated to DoD civilian employee job titles, but Bradshaw and her deputy, Marilee Fitzgerald, believed that DoD had civilians who were both willing and qualified to deploy for these positions. Working with the Joint Staff and the human resource directors for each of the services, CPP once again helped recruit, select, and deploy DoD civilians for overseas wartime assignments.[22] As with PRTs, relatively small numbers of DoD civilians deployed to advise Iraqi ministries, but these positions became enduring requirements. By 2009, these advisory requirements had become the basis for the Ministry of Defense Advisors program, a ministerial capacity-building effort that deploys senior DoD officials

---

[21] Jessie Riposo, Irv Blickstein, Stephanie Young, Geoffrey McGovern, and Brian McInnis, *A Methodology for Implementing the Department of Defense's Current In-Sourcing Policy*, Santa Monica, Calif: RAND Corporation, TR-944-NAVY, 2011.

[22] Interview with researchers conducting a related study at National Defense University's Center for Complex Operations, June 16, 2014.

to exchange expertise with their foreign counterparts in similar defense specialties.[23]

The third major force that increased the demand for DoD expeditionary civilians was U.S. Central Command's (USCENTCOM's) "requests" for civilians through the Request for Forces (RFF) and Global Force Management (GFM) processes. While the RFF process is designed to provide *capabilities* to combatant commanders, it offers a "civilian acceptable" option, which was increasingly used from 2007 to 2009. Also used more frequently during this period was the Joint Individual Augmentation (JIA) portion of the GFM process, due to the aforementioned operational tempo, the expansion of military missions, and the desire to work around force management levels.[24]

This increased demand for civilians was not matched by an addition of supply, however. In fact, according to one senior official in the Office of the Secretary of Defense (OSD), the Department of the Army (which was designated as the executive agent to fill joint billets in Iraq because it owned the ground mission and had the largest available force) "could not fill the positions."[25] Senior OSD officials and combatant command (CCMD) representatives asserted that this problem existed for three reasons: (1) the Army "did not make a real attempt to recruit skills or mission-specific competencies," (2) civilians in the other services "were not incentivized to deploy," and (3) "the services don't want to give up their civilians to joint missions."[26]

Collectively, these increases in demand and curtailments in the supply of DoD civilians led OUSD(P&R) to transform how DoD

---

[23] Jack Kem, "The Ministry of Defense Advisor (MoDA) Program: MoDA as a Component of the Civilian Expeditionary Workforce," in *From Mission Creep to Mission Gallop: Changing the Dynamics in Afghanistan*, NATO Training Mission–Afghanistan, 2011; Defense Security Cooperation Agency, "Ministry of Defense Advisors," web page, undated.

[24] Interviews with DoD, CCMD, and U.S. military officials, May–November 2014. Details on the RFF and GFM processes are provided in Defense Logistics Agency Instruction 3000.03, *Global Force Management*, January 4, 2013, and U.S. Joint Chiefs of Staff, *Joint Operational Planning*, Joint Publication 5-0, Washington, D.C., August 11, 2011b. The GFM process is discussed in greater detail in Chapter Three of this report.

[25] Interview with a DoD official, June 2014.

[26] Interviews with DoD, CCMD, and U.S. military officials, May–November 2014.

gained access to and deployed its civilian workforce. This transformation was codified in the January 2009 DoDD 1404.10, which established the Civilian Expeditionary Workforce (CEW) and made the management of CEW "sourcing" an OSD responsibility.[27] As an element of CPP, the CEW Program Office (CEWPO) assumed an operational function, and for nearly five years it recruited DoD civilians and civilians from the private sector, bringing them into DoD under temporary hiring authorities in an attempt to create a trained and ready cadre of civilians to perform jobs in combat zones. In fact, during that time, the CEWPO was responsible for recruiting, training, and deploying nearly 5,000 civilians into a variety of theaters across a wide range of job requirements.[28] However, at the same time, the CEWPO was responsible for creating policy to help guide the use of expeditionary civilians. This dual role proved challenging, as did the attempt to transform the way DoD managed its civilian workforce, and, ultimately, CPP terminated its plans to build a standing cadre and ceased its role in sourcing expeditionary civilians. The policy laid out in DoDD 1404.10, as well as its implementation, is explored in further detail in Chapter Two; Chapter Three elaborates with details of the evolution of the sourcing process for filling expeditionary civilian positions.

Objectives of This Study

In light of this history surrounding the deployment of DoD expeditionary civilians, the CEWPO within CPP's Defense Civilian Personnel Advisory Service asked RAND to assess its recent restructuring of DoD expeditionary civilian capability management. Specifically, CPP requested that RAND assess its decisions to (1) focus solely on its policy role with regard to expeditionary civilians, (2) have the military services assume operational responsibility for DoD expeditionary civil-

---

27 U.S. Department of Defense Directive 1404.10, *DoD Civilian Expeditionary Workforce*, January 23, 2009.

28 Interview with researchers conducting a related study at National Defense University's Center for Complex Operations, June 16, 2014.

ians, and (3) shift away from the large-scale CEW "cadre" model envisioned in DoDD 1404.10 (2009) to a more reactive model designed to respond to requirements for expeditionary civilians as they arise.

To address these issues, this study had two formal objectives: to conduct an end-to-end review of guidance that oversees the processes associated with deploying DoD civilians and to recommend guidelines for establishing and maintaining a civilian deployment capability to meet CCMD requirements. Three research questions are associated with these objectives:

1. What are the goals of DoD's civilian deployment capability, and what gaps currently exist between policy and practice pertaining to this capability?
2. What are the various CCMDs' perspectives regarding the utilization of expeditionary civilians? What are their requirements and plans for the use of expeditionary civilians in contingency operations, and how do these translate into manpower and skill requirements for deployable DoD civilians?
3. To what extent can we derive lessons from alternative civilian deployment models that could usefully be applied to a DoD model for deploying civilians? What are the most useful lessons from analogous cases?

It is critical to note that in pursuing these objectives and seeking to answer these questions, we aimed to take a *prospective, forward-looking* view of the possibilities for a civilian deployment framework that will continue to be viable ten to 20 years in the future, rather than a retrospective view examining the issues surrounding the implementation of the CEW and associated policy thus far. Therefore, while some degree of retrospection is necessary to understand the present context and future prospects for the successful deployment of expeditionary civilians under certain conditions, the overall intent of this report is to be forward-looking and to provide robust recommendations that will be of use to DoD as it moves forward with the deployment of civilians.

In looking toward the future, we should also note that the problem that the deployment of expeditionary civilians is meant to solve is that the U.S. government currently lacks deployable capacity to support complex contingency operations with the skill sets in which DoD civilians are qualified, and in which these personnel may enjoy a comparative advantage over other types of personnel. This is particularly true for civilian requirements not already assigned to a force, such as joint requirements labeled as "civilian acceptable." This lack of capacity is exacerbated by ongoing efforts to downsize the U.S. military and insource functions formerly reserved for contingency-support contractors. However, the nature, scale, and location of future potential contingency operations are currently unknown. In the midst of such uncertainty, the CEW construct was developed to provide a scalable potential source of such individuals.

It is similarly critical to note that during the course of this research, we encountered some difference of opinion among our interviewees regarding whether DoD civilians should be deployable at all. Thirty of our interviewees addressed this issue. Nineteen officials argued that the capability to deploy civilians should exist, four officials argued quite vehemently that civilians should not be deployed under any circumstances, and seven officials had mixed views on the issue.[29] Those arguing against DoD deployment of civilians felt that civilians were not prepared for deployments to the same extent as military personnel, that these are positions that can and should be filled by uniformed military (and the military should be grown as needed to fill these requirements), that any necessary deployable civilian positions would be more appropriately filled by DoS personnel, and that civilians can be a security risk in hostile environments—so much so that they may, in fact, require additional security for protection, thus serving to burden rather than support the uniformed military.

In addition to these differing perspectives on this issue, existing policy leaves open the question of whether civilians should be deployable, and if so, under what circumstances. In response to guidance from our research sponsor, this report proceeds from the assumption that

[29] Interviews with DoD, CCMD, and U.S. military officials, May–November 2014.

DoD aims to continue deploying civilians and believes that civilians are deployable (at least in certain situations). However, it is worth noting that there is some underlying debate throughout the defense community on this issue. Our discussion of Total Force policy in Chapter Two elaborates on this gap between policy and practice and points toward some potential remedies that may aid in resolving this debate.

Study Approach

To answer these questions, we worked with DoD officials to identify the goals of DoD's civilian deployment capability and any potential gaps between the current policies and practices pertaining to this program. We also reviewed CCMD requirements for expeditionary civilians, as well as the various CCMDs' perspectives with regard to the deployment of civilians, to establish baseline manpower and personnel management requirements for this capability to support contingency program management over time. We then drew on relevant lessons from comparative cases of civilian deployment policies and practice from both U.S. government organizations and foreign governments. From these comparative cases, we derived best practices and created a typology of four models of civilian deployment, highlighting the benefits and drawbacks of each. Combining the lessons of this typology with the policy analysis and survey of CCMDs' need for—and understanding of—expeditionary civilian capabilities, we developed an overall assessment of the viability of the current civilian deployment process. We then devised recommendations for establishing and maintaining a civilian deployment capability that could feasibly be used to meet CCMD requirements over the next several decades.

Much of the data for these tasks were collected through interviews with DoD and CCMD officials, as well as a review of relevant policy guidance, analysis, and assessments by such organizations as GAO and in the secondary literature. Over the course of the study, we interviewed a total of 83 individuals spanning 45 offices across DoD, other U.S. government agencies, and foreign governments (see Table 1.1). These included multiple directorates under OSD, the military services, and

Table 1.1
Organizations Interviewed for This Study

Agency Type	Office
U.S. government agencies	
DoD	Office of the Deputy Assistant Secretary of Defense for Civilian Personnel Policy
	Office of the Deputy Assistant Secretary of Defense for Total Force Management
	Office of the Under Secretary of Defense for Intelligence
	Office of the Deputy Assistant Secretary of Defense for Readiness
	Office of the Under Secretary of Defense for Policy
	Office of the Deputy Assistant Secretary of Defense for Strategy and Force Development
	Office of the Under Secretary of Defense for Acquisition, Technology, and Logistics (Operational Contract Support)
	U.S. Forces–Afghanistan
Military services	U.S. Army
	U.S. Army Corps of Engineers
	U.S. Navy
	U.S. Air Force
CCMDs	U.S. Central Command (USCENTCOM)
	U.S. Southern Command (USSOUTHCOM)
	U.S. European Command (USEUCOM)
	U.S. Africa Command (USAFRICOM)
	U.S. Pacific Command (USPACOM)
	U.S. Northern Command (USNORTHCOM)
	U.S. Special Operations Command (USSOCOM)
DoS	Bureau of Conflict and Stabilization Operations
	Bureau of South and Central Asian Affairs
	Bureau of Diplomatic Security
	Afghanistan and Pakistan Strategic Partnership Office (APSPO)

Table 1.1—Continued

Agency Type	Office
U.S. government agencies (cont.)	
USAID	Crisis Surge Support Staff (CS3) Office of Transition Initiatives (OTI)
U.S. Department of Homeland Security	Office of International Affairs U.S. Customs and Border Protection (CBP) Federal Emergency Management Agency (FEMA)
U.S. Department of Justice	Drug Enforcement Administration (DEA)
DoD Fourth Estate	Defense Intelligence Agency (DIA) Defense Logistics Agency (DLA) Defense Contract Management Agency (DCMA)
Foreign government agencies	
UK Ministry of Defence (MOD)	Support to Operations (S2O)
Canadian Department of National Defence	J1 (personnel) and human resources
Australian Department of Defence	Defence Science and Technology Organisation (DSTO, now the Defence Science and Technology Group)
European External Action Service (EEAS)	Civilian Planning and Conduct Capability (CPCC), Civilian Response Teams

each of the geographic CCMDs (USCENTCOM, USSOUTHCOM, USEUCOM, USAFRICOM, USPACOM, and USNORTHCOM), in addition to one functional CCMD (USSOCOM). To help us identify best practices, interviews included discussions with representatives of several DoD "Fourth Estate" agencies that have their own well-established civilian deployment programs, such as DLA, DIA, and DCMA. Other domestic organizations with which we conferred included the Bureau of Conflict Stability and Bureau of Diplomatic Security in DoS; the Office of International Affairs, CBP, DEA, and FEMA in the U.S. Department of Homeland Security; and USAID's

CS3 and OTI. Outside of the United States, we spoke with officials from the UK MOD's S2O team; the Canadian Department of National Defence; the Australian DSTO (now known as the Defence Science and Technology Group); and the European Union's EEAS CPCC.[30]

This research was conducted in 2014, and the findings were current as of mid-2015.

Organization of This Report

The remainder of this report begins with a gap analysis in Chapter Two detailing our findings on the policy context underlying current DoD efforts to deploy civilians, and the extent to which policy correlates with practice. Chapter Three details our findings on the operational aspects of DoD civilian deployment, including the evolution of sourcing processes, the integration of expeditionary civilians into manpower planning processes (such as the GFM process), and anticipated need for, and perspectives on, expeditionary civilian capabilities across the CCMDs. Chapter Four describes our comparative case analysis, delineates best practices from non-DoD organizations, and developes a typology of deployment models highlighting the benefits and drawbacks of various civilian deployment practices. Concluding the report, Chapter Five builds on the analysis in previous chapters to assess the viability of the current practice of DoD civilian deployment and provide recommendations to ensure that DoD expeditionary civilian capabilities are available to meet CCMD requirements over the long term.

---

[30] These interviews are attributed anonymously throughout this report in compliance with the U.S. Federal Policy for the Protection of Human Subjects (also known as the Common Rule). Both RAND's Institutional Review Board and human-subjects protection reviewers in DoD approved of this research method for this study. Organizational affiliation is included in the citation for each anonymous interviewee to give a sense of the individual's background and experience, but it should be noted that interviewees were not asked to represent their organizations in a confidential way. While interviewees were asked to respond based on their professional experiences, they were, in all cases, speaking for themselves rather than for their organizations in an official capacity.

CHAPTER TWO

Civilian Deployment in Policy Versus Practice

As noted in Chapter One, one of the primary tasks of this study was to perform an end-to-end review of the policies and procedures associated with DoD's civilian deployment capability. To do so, we performed a gap analysis of the discrepancies between policy and practice, utilizing both policy documents and interviews to determine (1) what existing policy says regarding DoD civilian deployment, (2) what existing DoD civilian deployment-related policies appear to be missing, and (3) what the gaps in policy, when compared with actual procedures in practice, are the key strategic considerations to take into account when designing a workable civilian deployment model. Our interviews, in particular, highlighted the following as critical issues for consideration:

- the larger context of Total Force management
- cost-efficiency
- recruitment, retention, and development of the DoD civilian workforce
- ability to meet operational needs.

In exploring these critical issues further, it became clear that several bodies of policy are particularly relevant to the topic of DoD civilian deployment. In addition to examining existing DoD policy on expeditionary civilians, we therefore assessed the literature and policy guidance on both Total Force management and strategic human capital planning (SHCP).

This chapter explores each of these three policy areas in turn, summarizing existing policy and contextualizing current DoD civilian deployment practice within the requirements of that policy. The analysis that follows highlights both strengths and weaknesses in existing policy and practice, suggesting potential areas for improvement.

DoD Policy on "Expeditionary Civilians"

The 2009 DoDD 1404.10, aptly titled *DoD Civilian Expeditionary Workforce*, was the reigning DoD policy on civilian deployment during the period in which we conducted our research. DoDD 1404.10 states that it is DoD policy to

> Rely on a mix of capable military members and DoD civilian employees to meet DoD global national security mission requirements. DoD civilian employees are an integral part of the Total Force. They serve in a variety of positions, provide essential capabilities and, where appropriate for civilians to do so, support mission requirements such as combat, contingencies, emergency operations; humanitarian and civic assistance activities; disaster relief; restoration of order; drug interdiction; and stability operations of the Department of Defense, herein collectively referred to as "expeditionary requirements."[1]

This guidance was an outgrowth of the 2007 National Defense Authorization Act (NDAA) requirement for a deployable cadre to support contingency program management.[2] DoDD 1404.10 aimed to achieve this policy goal through the creation of a preidentified, standing cadre of DoD employees who would be "organized, ready, trained, cleared, and equipped in a manner that enhances their availability to mobilize and respond urgently to expeditionary requirements." Collectively, this cadre was titled the DoD CEW and was envisioned to

---

[1] DoDD 1404.10, 2009, p. 2.

[2] Pub. L. 109-364, National Defense Authorization Act for Fiscal Year 2007, Section 854, October 17, 2006; interviews with DoD officials, April 30, 2014.

ultimately consist of between 20,000 and 30,000 DoD employees.[3] The CEW covered a wide variety of job specialties and positions, including but not limited to

- contracting staff
- security administration staff
- public affairs staff conducting media relations and developing communication strategies for deployed U.S. forces
- foreign affairs staff
- general attorneys
- transportation specialists performing vehicle maintenance
- information technology (IT) managers
- language specialists serving as translators and cultural advisers to U.S. officers
- intelligence specialists providing technical and IT support to intelligence personnel on U.S. bases in theater
- human resources assistance staff
- civil engineers working with host-nation and U.S. contractors and nongovernmental organizations on civil-military operations projects
- personnel in the Afghanistan-Pakistan (AfPak) Hands program performing development and government capacity-building work in Afghanistan[4]
- personnel in the Ministry of Defense Advisors program performing defense capacity-building activities.[5]

---

[3] DoDD 1404.10, 2009; Molly Dunigan, *Considerations for the Civilian Expeditionary Workforce: Preparing to Operate Amidst Private Security Contractors*, Santa Monica, Calif.: RAND Corporation, OP-359-OSD, 2012, p. 1.

[4] AfPak Hands is an "all-in" language and cultural immersion initiative developed in 2009 under Chairman of the Joint Chiefs of Staff Admiral Mike Mullen to help the International Security Assistance Force accelerate the continual transition of more responsibility to Afghanistan's government and security forces. See Matthew Chlosta, "'AfPak Hands' Begin Immersion Training," American Forces Press Service, May 5, 2010. These positions, while managed for some time by the CEW, required a more extensive commitment, cultural immersion, and more frequent contact with Afghan officials than did typical CEW positions. See Dunigan, 2012, pp. 2–3.

[5] Interviews with DoD officials, April 30, 2014; Dunigan, 2012, pp. 2–3; DoD, 2011.

There were several structural changes pertaining to DoD civilian workforce management after the issuance of this 2009 guidance. The CEWPO was established within the International/Expeditionary Support (IES) Office under CPP, along with a separate director to manage both the policy and operational aspects of the CEW.[6] Moreover, the 2009 guidance designated three main categories of personnel:

1. Emergency-Essential (E-E): "a position-based designation to support the success of combat operations or the availability of combat-essential systems"
2. Non-Combat Essential (NCE): "a position-based designation to support the expeditionary requirements in other than combat or combat support situations"[7]
3. Capability-Based Volunteer (CBV): "an employee who may be asked to volunteer for deployment, to remain behind after other civilians have evacuated, or to backfill other DoD civilians who have deployed to meet expeditionary requirements in order to ensure that critical expeditionary requirements that may fall outside or within the scope of an individual's position are fulfilled."[8]

The directive also created the "Capability-Based Former Employee Volunteer Corps," which was to consist of "a collective group of former (including retired) DoD civilian employees who have agreed to be listed in a database as individuals who may be interested in returning to Federal service as a time-limited employee to serve expeditionary

[6] Interviews with DoD officials, April 30, 2014.

[7] Both E-E and NCE personnel were also to be designated as "key employees" in accordance with DoDD 1200.7. DoDD 1200.7 defines key employees as occupying "key positions," which are "Any Federal position that shall not be vacated during a national emergency or mobilization without SERIOUSLY impairing the capability of the parent Federal Agency or office to function effectively" (emphasis in original). U.S. Department of Defense Directive 1200.7, *Screening the Ready Reserve*, Enclosure 2, November 18, 1999, certified current as of November 21, 2003, p. 7.

[8] DoDD 1404.10, 2009, p. 3.

requirements or who can backfill for those serving other expeditionary requirements."[9]

However, as noted in Chapter One, the vision for the CEW never fully came to fruition. As a result, the IES Office within CPP was working to revise policy related to the CEW at the time of this writing. CPP has also opted to forgo a large-scale cadre of preidentified employees for deployment in favor of a smaller force of individuals totaling approximately 5,300 across DoD. In that process, the term *civilian expeditionary workforce*, or *CEW*, has been replaced by *expeditionary civilians*, which is intended to convey the replacement of a large, standby cadre with a smaller group of individuals sourced primarily as requirements arise.[10]

Against this background, we sought to determine how closely DoD's policies on expeditionary civilians correlate with civilian deployment practices and experiences. The goal of DoD's civilian deployment capability, as written in the 2007 NDAA, is to "support contingency program management."[11] DoDD 1404.10 highlights that expeditionary civilians are intended, where appropriate, to "support mission requirements such as combat, contingencies, emergency operations, humanitarian and civic assistance activities, disaster relief, restoration of order, drug interdiction, and stability operations of the Department of Defense."[12] Meanwhile, the Commission on Wartime Contracting in Iraq and Afghanistan recommended developing a deployable civilian cadre for acquisition management and contractor oversight "to reduce over-reliance on contractors."[13] Interviews with various DoD and CCMD officials indicate that the initial impetus to utilize expe-

---

9 DoDD 1404.10, 2009, p. 3.

10 Interviews with DoD officials, October 16, 2014.

11 Pub. L. 109-364, National Defense Authorization Act for Fiscal Year 2007, Section 854, October 17, 2006.

12 DoDD 1404.10, 2009.

13 Commission on Wartime Contracting in Iraq and Afghanistan, *Transforming Wartime Contracting: Controlling Costs, Reducing Risks*, final report to Congress, August 2011.

ditionary civilians was to fill low-density, high-demand positions with personnel with skills not easily found in the uniformed military.[14]

Interestingly, the use of expeditionary civilian personnel in practice originated from a desire to reduce stress on the uniformed military. Quantitative data collected from the CCMDs and interviews with DoD and CCMD officials suggest that, since 2009, the bulk of the DoD expeditionary civilian workforce has been performing jobs that could be performed by the military (i.e., *not* low-density, high-demand positions).[15] Interviews indicate that expeditionary civilians have been used in recent years, for the most part, to circumvent force management levels that preclude the hiring of additional uniformed military personnel in theater. Fourteen interviewees noted this phenomenon, spanning personnel from USCENTCOM, USSOUTHCOM, USSOCOM, the U.S. Air Force, the U.S. Navy, and the Joint Staff.[16] For instance, one CCMD interviewee noted, "Force management levels have a huge effect," leading CCMDs to utilize civilians.[17] Another CCMD interviewee noted, "Over the past couple of years, force management levels have made us start looking for civilians just to fill positions."[18] While force management levels presented a quantifiable and easily identified challenge that expeditionary civilians were used to surmount, the overarching narrative here is that the U.S. government simply did not have sufficient deployable capacity to meet its needs during this period. The fact that civilians and military personnel are "counted" differently in many contexts but may be able to perform similar tasks speaks to the need to fully define civilian deployment in the context of the Total Force.

---

[14] Correspondence with a CCMD official, June 16, 2014.

[15] Correspondence with a CCMD official, June 16, 2014; interview with a CCMD official, September 9, 2014; "J-1 Spreadsheets for CCMD Requirements for Deployable DoD Civilians," provided by CEWPO to RAND in April 2014.

[16] Interviews with DoD, CCMD, and U.S. military officials, May–November 2014.

[17] Interview with a CCMD official, June 17, 2014.

[18] Interview with a CCMD official, August 28, 2014.

Total Force Management

The notion of the Total Force holds different meanings in different contexts. It has evolved over time, and its implementation remains uneven across the various DoD components and offices. The concept originated in policy on August 21, 1970, when Secretary of Defense Melvin Laird dispatched a memorandum coining the term. Known as the Laird Memorandum, this document stressed the need to capitalize on all available forces:

> Within the Department of Defense, these economies will require reductions in overall strengths and capabilities of the active forces, and increased reliance on the combat and combat support units of the Guard and Reserves. . . . Emphasis will be given to concurrent considerations of the *total forces*, active and reserve, to determine the most advantageous mix to support national strategy and meet the threat. A *total force concept* will be applied in all aspects of planning, programming, manning, equipping and employing Guard and Reserve forces. [Emphasis added][19]

The original intent of the Total Force concept was quite broad and included allied and host-nation forces in occupied areas, as well as U.S. forces. As conceived by Secretary Laird, Total Force policy was to apply equally to U.S. forces, both active and reserve, and North Atlantic Treaty Organization (NATO) and other allied forces. It also applied to South Vietnamese forces that were to assume greater responsibility for the defense of their country under the Nixon policy of "Vietnamization." According to the tenets of the Total Force concept, America's NATO allies would be required to share a larger portion of the burden of defending the European continent, South Vietnamese soldiers would be forced to fight the war in Southeast Asia without the assistance of U.S. ground troops, and less expensive U.S. reserve forces would be expected to assume more missions previously assigned

---

[19] Melvin O. Laird, Secretary of Defense, "Readiness of the Selected Reserve," memorandum, August 21, 1970, pp. 1–2, quoted in Patrick M. Cronin, *The Total Force Policy in Historical Perspective*, Alexandria, Va.: Center for Naval Analyses, June 1987.

to U.S. active forces. In addition, the Laird Memorandum emphasized that the reserves were to be the "initial and primary source of augmentation of the active forces in any future emergency requiring a rapid and substantial expansion of the active forces."[20]

However, in subsequent years Total Force policy came to be associated almost exclusively with the integration of the reserve components into the active military services. This interpretation of the Total Force is still quite prevalent today in the services' use of the term. In the aftermath of the Vietnam War, U.S. Army Chief of Staff General Creighton Abrams adopted Total Force policy to treat the three components of the Army—the active component, Army Reserve, and Army National Guard—as a single force.[21] This concept has carried through to the present, with Secretary of the Army John M. McHugh signing the Army Total Force Policy in September 2012 to align the Army with DoD guidance requiring the military services to manage their reserve components as an operational force.[22] The U.S. Air Force conceives of the term similarly, defining it as consisting of three components—the Regular Air Force, Air National Guard, and Air Force Reserve.[23] The Navy diverges slightly from the other services, defining the Total Force as "active and reserve Sailors and Navy civilians."[24] This inclusion of DoD civilians in the Total Force is in line with recent DoD policy on the topic. For instance, the 2006 Quadrennial Defense Review (QDR) placed renewed emphasis on the management of the Total Force, which it defined as consisting of military, civilian, and contractor personnel:

---

[20] Cronin, 1987, p. 7.

[21] Army National Guard, "Constitutional Charter of the Guard," 2009.

[22] Association of the U.S. Army, "Army Total Force Policy: Fully Integrating the Operational Reserve," *Defense Report*, Arlington, Va.: Institute of Land Warfare, December 2012

[23] U.S. Senate Committee on Armed Services, "Department of Defense Authorization of Appropriations for Fiscal Year 2015 and the Future Years Defense Program: The National Commission on the Structure of the Air Force," hearing, Washington, D.C., April 29, 2014.

[24] U.S. Navy, *Navy's Total Force Vision for the 21st Century*, January 2010, p. 1.

> The Department's Total Force—its active and reserve military components, its civil servants, and its contractors—constitutes its warfighting capability and capacity. . . . The Total Force must continue to adapt to different operating environments, develop new skills and rebalance its capabilities and people if it is to remain prepared for the new challenges of an uncertain future. . . . Increasing the adaptability of the Total Force while also reducing stress on military personnel and their families is a top priority for the Department."[25]

This emphasis on the tripartite division of the Total Force into military, civilian, and contractor components is significant. It aligns with the definition of *Total Force* found in Title 10 of the U.S. Code (U.S.C.), Section 129a, which states, "The Secretary of Defense shall establish policies and procedures for determining the most appropriate and cost efficient mix of military [active and reserve], civilian, and contractor personnel to perform the mission of the Department of Defense."[26] It is also reiterated in the 2010 QDR, which, as Kathleen Hicks, Deputy Under Secretary of Defense for Strategy, Plans, and Forces, noted at a 2010 hearing of the Commission on Wartime Contracting in Iraq and Afghanistan, was "the first QDR to elevate the need to preserve and enhance the force to a priority objective." Significantly, Hicks highlighted the QDR's emphasis on the CEW as part of the Total Force and the fact that the overall strategy must be implemented to be complete:

> The department must ensure the long-term viability of the all-volunteer force, the nation's most precious military resource. This will require policies that sustain the rotation base, provide care for our people, and adapt as required to the environment. The 2010 QDR specifically calls for the department to further assess whether we have the right combination of military and DoD civilian personnel and contracted services. The QDR report also

---

[25] U.S. Department of Defense, *Quadrennial Defense Review Report*, February 6, 2006, p. 75.

[26] Title 10 U.S. Code, Subtitle A, Chapter 3, Section 129a, "Armed Forces: General Policy for Total Force Management," p. 111.

> highlighted several initiatives in total force management, including *the civilian expeditionary workforce*, in-sourcing and [operational contract support] itself. The QDR's primary purpose is to set the department on a long-term strategic course. The strategy is incomplete without execution, however. [Emphasis added][27]

Management of the Total Force, conceived as consisting of military, civilian, and contractor personnel, is a critical prerequisite to the successful implementation of DoD civilian deployment policy. Each of these respective workforces offers different features and capabilities, and underlying the notion of the Total Force is the idea that the relative sizes of these workforces are scalable and should ideally be tailored to the contingency at hand (see Figure 2.1).

Figure 2.1
The Total Force Concept as a Set of Overlapping, Scalable Workforces Offering Distinct Capabilities

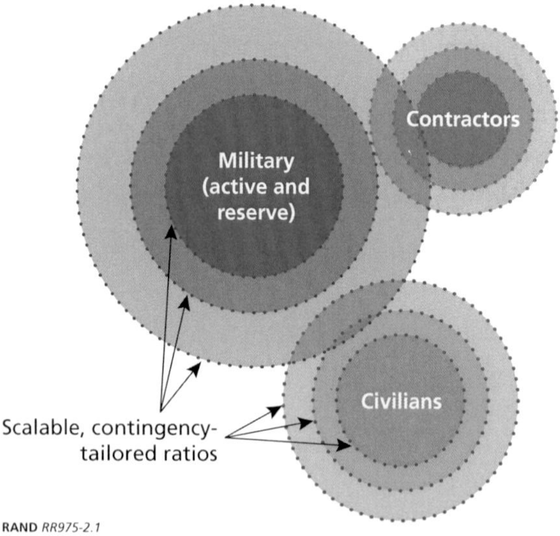

RAND *RR975-2.1*

[27] Kathleen Hicks, testimony before the Commission on Wartime Contracting at the hearing "Total Force Policy, the QDR, and Other Defense and Operational Planning: Why Does Planning for Contractors Continue to Lag?" Washington, D.C., July 12, 2010.

However, as noted, the implementation of Total Force policy, at least insofar as it applies to civilians and contractors, has been uneven. The Joint Staff and most CCMDs are still largely focused on military personnel requirements, and interviews with OSD officials indicated that meetings of the Global Force Management Board—which manages the force allocation process—generally focus on major components of force structure, not civilian deployments.[28]

More significantly for the fate of the DoD expeditionary civilian concept, existing guidance related to Total Force management leaves some question as to whether civilians should be deployable at all—and, if so, under what circumstances. Several specific pieces of guidance are relevant here. First, a 2003 memo by David Chu, then–Under Secretary of Defense for Personnel and Readiness, stated, "As a rule, civilian knowledge and facility shall be developed to engender a 'profession of governance'; whereas military knowledge and skills shall be developed to engender a 'profession of arms.'"[29] This distinction raises the question of what deployed positions would allow civilians to maintain their "profession of governance" and not tread on the "profession of arms"—a question that has yet to be answered in existing policy on either expeditionary civilians specifically or Total Force management more broadly.

Second, a 2005 DoDD outlining guidance for manpower management stated,

> Manpower shall be designated as civilian except when military incumbency is required for reasons of law, command and control of crisis situations, combat readiness, or *esprit de corps*; when unusual working conditions are not conducive to civilian employment; or when military-unique knowledge and skills are required for successful performance of the duties.[30]

[28] Interview with a former senior DoD official, October 20, 2014.

[29] David S. C. Chu, Under Secretary of Defense for Personnel and Readiness, "Interim Policies and Procedures for Strategic Manpower Planning and Development of Manpower Estimates," memorandum, December 10, 2003.

[30] U.S. Department of Defense Directive 1100.4, *Guidance for Manpower Management*, February 12, 2005.

Such language leads one to wonder what exactly constitutes "unusual working conditions not conducive to civilian employment," and whether this would include any or all deployment scenarios, particularly those involving hostile environments. Again, this issue has yet to be clarified in existing policy.

More recent policy on when to utilize military personnel versus civilians versus contractors for various tasks raises a very similar question as to whether civilians should *ever* be deployable and, if so, in what circumstances. Issued in 2010, DoDI 1100.22, *Policies and Procedures for Determining Workforce Mix*, specifies the following criteria for determining when to use respective components of the Total Force:

> [E]ven if a function is not [inherently governmental] or exempted from private sector performance, it shall be designated for DoD civilian performance . . . unless an approved analysis for either of the following exceptions has been addressed consistent with the DoD Component's regulatory guidelines:
>
> (1) a cost comparison or a public-private competition . . . shows that DoD civilian personnel are not the low-cost provider.
>
> (2) There is a legal, regulatory, or procedural impediment to using DoD civilian personnel. This shall include determinations by Human Resources (HR) officials that DoD civilians cannot be hired in time, or retained to perform the work. . . .
>
> *[M]anpower shall be designated as civilian except when one or more of the following conditions apply:*
>
> (1) Military-unique knowledge and skills are required for performance of the duties.
>
> (2) Military incumbency is required by law, [executive order], treaty, or international agreement.
>
> (3) Military performance is required for command and control, risk mitigation, or *esprit de corps*.

(4) *Military manpower is needed to provide for overseas and sea-to-shore rotation, career development, or wartime assignments.*

(5) Unusual working conditions or costs are not conducive to civilian employment. [Emphasis added][31]

This guidance again highlights the need to clarify the definition of "unusual working conditions" that are "not conducive to civilian employment" but also raises the additional perplexing question of whether it is *only* military manpower that is allowed for overseas and sea-to-shore rotation and wartime assignments. If so, this would be a clear restriction against the deployment of expeditionary civilians. While it is unlikely that DoD intended to write policy restricting civilian deployment, the key takeaway here is that existing DoD policy does not reflect a consistent, clear philosophy on whether or to what extent to deploy civilians.

Three other issues related to management of the Total Force are relevant to both the motivation for and long-term feasibility of DoD's expeditionary civilian capability: civilian ability to work across the joint workforce, insourcing of numerous capabilities previously outsourced to contractors, and the cost-efficiency of deploying civilians as opposed to military personnel or contractors for various tasks. Each of these is considered in turn in the following sections.

Civilian Ability to Work Across the Joint Workforce

In 1986, Congress passed the Goldwater-Nichols Department of Defense Reorganization Act, which made the most sweeping reforms to DoD since it was established in 1947. Interservice rivalries, which had long prevented the U.S. military from operating as a joint force, served as the impetus for these reforms. Among the reforms were significant changes to the personnel management of military officers (mandating joint duty assignments and joint professional military education as requirements for promotion) and the transfer of operational control of service units to joint commanders in the functional and geographic

---

[31] U.S. Department of Defense Instruction 1100.22, *Policy and Procedures for Determining Workforce Mix*, April 12, 2010, pp. 2–3.

CCMDs.[32] Despite strong opposition from the services, the passage of Goldwater-Nichols helped make possible the operational successes that the U.S. military enjoyed in the years that followed.[33]

At the time of Goldwater-Nichols, DoD civilians were rarely used in an expeditionary manner. Therefore, Congress had little incentive or justification to reform the organizational structure and personnel management of the DoD civilian workforce. Twenty years later, however, OIF and OEF exposed the fact that much of the same service-specific parochialism that had plagued the uniformed military prior to the passage of Goldwater-Nichols still applied to the use of service-owned civilians. Not surprisingly, several interviewees suggested that a joint civilian workforce would create positive incentives to both motivate civilians to deploy and encourage their home offices (i.e., the offices for which they work when not deployed) to support their deployments, and it could also increase CCMD utilization of DoD expeditionary civilians.

According to USSOCOM officials, "joint civilians" would be beneficial for USSOCOM because they would allow the commander to globally manage the civilians who work under the CCMD. This sentiment was motivated, in part, by the fact that civilians working for USSOCOM are "owned" by an individual service. If an Air Force civilian stationed in Tampa, Florida, wanted to deploy, for example, he or she would have to apply for job as an Army civilian if the Army was the CCMD's executive agent.[34] According to one USSOCOM official, this has had two adverse effects: (1) "high-performing people were not let go by the parent organization," and (2) there were "no incentives for deploying."[35]

---

[32] Ronald Cole, "Grenada, Panama, and Haiti: Joint Operational Reform," *Joint Force Quarterly*, Fall–Winter 1998–1999.

[33] James Locher, *Victory on the Potomac: The Goldwater-Nichols Act Unifies the Pentagon*, Austin, Texas: Texas A&M University Press, 2004.

[34] Interviews with four USSOCOM officials, September 2014. See Chapter Three for a more comprehensive definition and discussion of executive agency and a list of the various services assigned as executive agents to each CCMD.

[35] Interview with a USSOCOM official, September 2014.

USAFRICOM officials offered similar perspectives and stated that joint requirements for civilian promotions would be beneficial because they would give civilians the ability to work across the services and for different CCMDs. This would be good for civilians filling either expeditionary positions for the CCMDs or nonexpeditionary positions across the services.[36]

One USEUCOM official asserted that any civilian assigned to a joint headquarters should be categorized as a "joint civilian," because civilians who switch service affiliation are disadvantaged by being precluded from applying to future positions in their original service, despite the fact that they may have gained the bulk of their expertise while a member of that service. In fact, this USEUCOM civilian had a 26-year career as an Air Force civilian, but to transfer to USEUCOM she had to become an Army civilian. She said this precluded her from applying for internal Air Force jobs in the future—a problem made worse by the five-year rotation policy outlined in DoDI 1400.25, Volume 1230.[37] Thus, the lack of "jointness" limited her future options.[38]

USPACOM and USNORTHCOM officials agreed that "joint civilians" would be beneficial regardless of whether they were employed for expeditionary purposes. Furthermore, both stated that the Joint Staff was aggressively pursuing reforms in this area such that it had even formed a working group to build a more joint-focused civilian workforce.[39]

Indeed, this is an active issue currently being addressed by the Joint Staff, which since July 2014 has hosted monthly video-teleconferences with CCMD J1 directorate representatives to address issues related to service executive agent support for civilian personnel.[40]

---

[36] Interviews with two USAFRICOM officials, September 2014.

[37] U.S. Department of Defense Instruction 1400.25, *DoD Civilian Personnel Management System, Volume 1230: Employment in Foreign Areas and Employee Return Rights*, July 26, 2012.

[38] Interview with a USEUCOM official, September 2014.

[39] Interviews with USPACOM and USNORTHCOM officials, September and November 2014.

[40] Correspondence with a Joint Staff official, November 6, 2014; interview with a CCMD official, November 5, 2014.

Notably, these problems are, for the most part, irrelevant to "CEW"-type requirements—for which one would *not* have to leave one's component or executive agent to support another CCMD's requirement. However, they are relevant to other situations in which DoD civilians want to apply for deployed CCMD positions but the CCMD in question has a different executive agent than the civilian's current service component. The prevalence of this problem was one of the drivers of DoDD 1404.10 and the creation of the CEW, and it illustrates the practical significance of having a CEW or some other form of expeditionary civilian workforce.

Insourcing

The practice of outsourcing defense functions to the private sector became increasingly popular in the 1990s and early 2000s against the backdrop of post–Cold War defense reductions, the Gulf War, and the Balkans conflicts. As a result, the DoD civilian workforce declined by about 38 percent between 1989 and 2002, dropping from 1,075,437 civilians to 670,166.[41] However, in the mid-2000s, several high-profile incidents revealed problems with contractor transparency, accountability, and overbilling in Iraq and Afghanistan and called such extensive outsourcing policies into question.[42] Following these incidents, the fiscal year (FY) 2008 NDAA added new language to 10 U.S.C. requiring the Office of the Under Secretary of Defense for Personnel and Readiness to develop guidelines and procedures to ensure that DoD considers using civilians to perform new functions or functions performed by contractors. Deputy Secretary of Defense Gordon England consequently signed a memorandum commonly referred to as the "Insourcing Memo" in April 2008, issuing these guidelines:

---

[41] U.S. General Accounting Office, *DoD Personnel: DoD Actions Needed to Strengthen Civilian Human Capital Strategic Planning and Integration with Military Personnel and Sourcing Decisions*, Washington, D.C., GAO-03-475, 2003.

[42] Riposo et al., 2011, p. 3; Walker, 2008, pp. 1, 9; James Glanz and Alissa J. Rubin, "From Errand to Fatal Shot to Hail of Fire to 17 Deaths," *New York Times*, October 3, 2007; Daniel Schorn, "Billions Wasted in Iraq?" *60 Minutes*, February 12, 2006; Erik Eckholm, "U.S. Contractor Found Guilty of $3 Million Fraud in Iraq," *New York Times*, March 10, 2006.

> The guidelines and procedures state that DoD Components can use DoD civilian employees to perform new functions or functions that are performed by a contractor if an economic analysis shows that DoD civilian employees are the low cost provider, or the DoD component has determined, consistent with DoD Instruction 1100.22 "Guidance for Determining Workforce Mix," that the function under review is inherently governmental or exempt from civilian performance. This guidance will help ensure that when DoD components make decisions to use DoD civilian employees, the decisions are fiscally informed and analytically based.[43]

Two days later, in his FY 2010 budget message, Secretary of Defense Robert Gates called for growing the civilian workforce by replacing contractors with DoD civilian personnel:

> Under this budget request, we will reduce the number of support-service contractors from our current 39 percent of the Pentagon workforce to the pre-2001 level of 26 percent, and replace them with full-time government employees. Our goal is to hire as many as 13,000 new civil servants in FY '10 to replace contractors and up to 30,000 new civil servants in place of contractors over the next five years.[44]

The Obama administration continued this emphasis on insourcing, issuing a number of executive branch initiatives as well as a May 2009 memorandum suggesting that "contractors may be performing inherently governmental functions."[45] The President's FY 2011 budget submissions reiterated concerns regarding an overreliance on contractors.[46] To meet these initiatives, the defense community has taken specific, measured steps—with, for instance, the Secretary of the

---

[43] England, 2008.

[44] U.S. Department of Defense, "DoD News Briefing with Secretary Gates from the Pentagon," transcript, April 6, 2008.

[45] Barack Obama, "Government Contracting," memorandum, March 4, 2009, p. 2.

[46] American Bar Association Public Contract Law Section, Legislative Coordinating Committee, Insourcing Initiatives, March 6, 2010, cited in Kate M. Manuel and Jack Maskell,

Army testifying in February 2010 that the Army intended to insource 7,162 positions in FY 2010 and 11,084 positions in FYs 2011–2015.[47]

Such specific goals have led to criticism that insourcing has become nothing more than a quota-driven exercise.[48] Moreover, the emphasis on insourcing has been trailed by an extensive legal debate, the details of which were outside of the scope of this study.[49] Yet, the insourcing debate is critical to overall discussions regarding the balance of various components of the Total Force in relation to one another. As a Defense Science Board task force noted in 2000, "Rapid downsizing during the last ten years" had been a catalyst for rethinking the balance between components of the Total Force—contractors, civilian person-

---

Insourcing Functions Performed by Federal Contractors: Legal Issues, Washington, D.C.: Congressional Research Service, February 22, 2013.

[47] Matthew Weigelt, "Army Vows to Cut 7,000 Contractor Jobs This Year," *Washington Technology*, February 23, 2010.

[48] See Manuel and Maskell, 2013, p. 3, citing the following: Professional Services Council, letter to the Secretary of Defense, May 3, 2010; National Defense Authorization Act for FY 2011 (which responded, in part, to such concerns by prohibiting DoD from establishing goals or quotas for insourcing functions), Pub. L. 111-383, Section 323, 124 Stat. 4184, January 7, 2011 (codified at 10 U.S.C. 2463); and National Defense Authorization Act for FY 2012, Pub. L. 112-81, Section 931(a), 125 Stat. 1543, December 31, 2011.

[49] Much of this debate has focused on what constitutes "inherently governmental" activities. The term *inherently governmental* dates to 1966, when the first Office of Management and Budget Circular No. A-76 was issued. It stated, "Certain functions are inherently governmental in nature, being so intimately related to the public interest as to mandate performance only by federal employees" (quoted in Acquisition Advisory Panel, *Report of the Acquisition Advisory Panel to the Office of Federal Procurement Policy and the United States Congress*, January 2007).

In addition to a prohibition on contractor performance of inherently governmental activities, there are three other major restrictions on the use of contractors in existing policy and law: (1) special rules about the use of contractors to perform functions closely associated with inherently governmental functions, (2) a prohibition on the use of personal-services contracts, and (3) DoD-specific exemptions from private-sector performance of specific commercial functions identified in law and policy. The underlying message behind each of these restrictions is straightforward: Only government employees should wield the authority of the government. However, due to the level of interpretation required to assess whether various functions meet these criteria, consistent implementation poses a challenge. See Riposo et al., 2011, p. x; and Manuel and Maskell, *Insourcing Functions Performed by Federal Contractors: An Overview of the Legal Issues*, Washington, D.C.: Congressional Research Service, May 5, 2011.

nel, and military personnel.[50] This has implications for both current and future deployments of DoD civilians as insourcing efforts create a strong demand for DoD civilians who can deploy to fill positions previously filled by contractors in conflict environments. Yet, as noted earlier, Total Force policy contains ambiguities regarding the appropriate extent to which civilians, as opposed to uniformed military personnel or contractors, should be *deployed*. The push to insource a large proportion of defense positions therefore serves to emphasize the significance of clarifying Total Force policy on whether, to what extent, and in what situations civilian deployment is appropriate.

Cost-Efficiency

Cost is paramount when considering whether to deploy civilians, and—to the extent that they should be deployed—how to structure and manage expeditionary civilian capabilities. Yet, our interviews revealed that the CEWPO was never fully resourced following the development of the original CEW concept and policy in DoDD 1404.10, and there are blatant inconsistencies in perceptions across DoD and the services/CCMDs regarding the relative costs of military versus civilian personnel. Thirteen out of 15 interviewees—spanning USCENTCOM, USSOUTHCOM, the Air Force, Army, Navy, and Joint Staff—noted that one reason *not* to utilize expeditionary civilians was that they were more expensive to deploy than military personnel.[51]

Yet, two interviewees, both from the same office within OSD, said the opposite—that an expeditionary civilian force was less expensive to deploy than military personnel.[52] This discrepancy poses a puzzle, and one that is critical to solve: Is it more or less cost-efficient to deploy civilians than military forces? The answer may very well depend on the scenario or position in question, in which case it would be useful to understand which positions are more cost-effective for civilian as opposed to military deployment. As we discuss later, such knowledge

---

[50] Defense Science Board, *Report of the Task Force on Human Resources Strategy*, Washington, D.C.: U.S. Department of Defense, February 2000.

[51] Interviews with Joint Staff, CCMD, and U.S. military officials, May–November 2014.

[52] Interviews with senior DoD officials, June 18, 2014.

would be extremely valuable in informing military and workforce planning.

However, we were unable to find any rigorous study to date directly comparing the relative costs of CEW personnel, or expeditionary civilians more broadly, to military personnel. We did find several studies assessing particular aspects of this problem, outlined later, but they provide little clarification as to the larger question of comparative civilian versus military deployment costs. Indeed, we were able to find only one small-scale analysis taking a comparative view of military versus civilian deployment costs, completed by the U.S. Air Force several years ago.

A 1998 RAND study sponsored by CPP used two cost analysis approaches to examine whether the transfer of military tasks to civil service personnel would entail cost savings. RAND found that one approach showed civil service personnel to be less costly than military personnel, while the other approach showed that civilian personnel were cost-effective at some civil service grades but not at others.[53] While this study provides information on the relative costs of civil service versus military personnel, it does not address civilian *deployment* or the costs of expeditionary civilians specifically.

Meanwhile, prior research does exist on the costs of deploying DoD and other U.S. government civilians, but not in relation to the uniformed military. When the civilian deployment concept was modeled as a standby cadre, the CEWPO analyzed the deployment, labor, and support costs of deploying 600 CEW personnel. However, it did not compare the costs of deploying other types of forces, and these estimates—as with any costing methodology—include numerous assumptions.[54] Moreover, because CPP is moving away from a standby

---

[53] Susan M. Gates and Albert A. Robbert, *Comparing the Costs of DoD Military and Civil Service Personnel*, Santa Monica, Calif.: RAND Corporation, MR-980-OSD, 1998.

[54] The cost estimates performed by CPP and the CEWPO found that human resources support costs would total $700,901; mobilization site, management support, and airfare costs would total $1,824,000; and CEWPO staff costs would total $937,882. The estimates assume unit deployment center costs per deployee of $1,500 and a CEWPO staff of seven full-time equivalents: six GS-13 through GS-15 employees, plus one GS-7 administrative assistant. Also built into these estimates is an assumption that only 10 percent of the

CEW and is structuring management of expeditionary civilian deployment differently at present, the assumptions made here would need to be updated to reflect the new direction of the concept.

The Special Inspector General for Afghanistan Reconstruction and the DoS Office of Inspector General have also estimated the costs of civilian deployment. In a joint 2011 publication, the two offices estimated that it costs the U.S. government between $425,926 and $570,998 to deploy a civilian employee to Afghanistan for one year.[55] This estimate applies specifically to U.S. civilian employees "deployed to Afghanistan under Chief of Mission authority" and is higher than the personnel costs estimated by the CEWPO in 2013 to deploy a GS-15 DoD civilian, which totaled $346,226.[56]

Meanwhile, there are separate analyses delineating the costs of deploying uniformed military personnel, but they too are devoid of any mention of the relative per-person costs of military deployment in relation to the costs of deploying other types of forces. The Center for Strategic and Budgetary Assessments estimated the cost per service member deployed, which ranged from $1 million to $2 million per year from FY 2008 to FY 2014 (see Figure 2.2).[57]

---

600 CEW personnel would require predeployment training at Camp Atterbury in Muscatatuck, Indiana, a "highly realistic . . . contemporary, and developmental testing environment" where much of the scenario-based training for joint, interagency, intergovernmental, multinational, and nongovernmental capabilities deploying to Iraq and Afghanistan has occurred. Together, these estimates add up to a total of $3,462,783 in support costs for a CEW of 600 deployees. Meanwhile, CPP estimated that labor costs for these 600 CEW deployees would total $108,462,783. This estimate is based on an average CEW salary of $175,000. Civilian Personnel Policy/Defense Civilian Personnel Advisory Service, CEW Program Office, "CEW Support Costs and FTE Breakdown Version 3," unpublished, provided to RAND in April 2014.

55 Office of the Special Inspector General for Afghanistan Reconstruction and Department of State Office of Inspector General, *The U.S. Civilian Uplift in Afghanistan Has Cost Nearly $2 Billion, and State Should Continue to Strengthen Its Management and Oversight of the Funds Transferred to Other Agencies*, SIGAR Audit-11-17 and State OIG Aud/SI-11-45 Civilian Uplift, September 8, 2011, Appendix II.

56 Julie Blanks, International/Expeditionary Support Office, *CEW Transition*, briefing, May 3, 2013, slide 16.

57 Todd Harrison, *Chaos and Uncertainty: The FY2014 Defense Budget and Beyond*, Washington, D.C.: Center for Strategic and Budgetary Assessments, October 24, 2013, p. 12.

Figure 2.2
Center for Strategic and Budgetary Assessment Estimates of Costs per Service Member Deployed, FY 2008–FY 2014

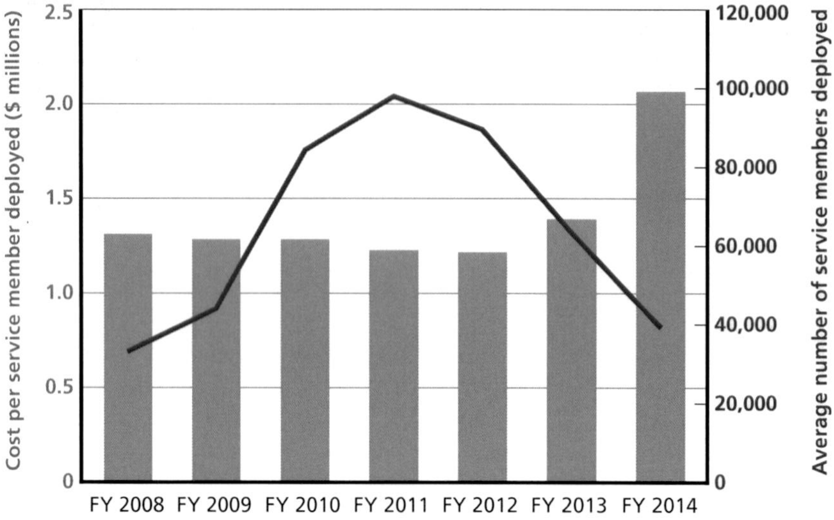

SOURCE: Harrison, 2013, p. 12.
RAND *RR975-2.2*

As noted earlier, the Air Force developed a comparative estimate of the relative costs of civilian versus military deployment—the only cost estimation analysis of its kind that we were able to find in our review of research on this topic. Air Force personnel devised this comparative estimate several years ago, examining the relative costs of military versus civilian personnel for a six-month deployment to OIF/OEF.[58]

---

[58] In creating this estimate, the Air Force made several key assumptions, including profiling the personnel whose costs were used to calculate the estimate. The military profile was of an Air Force major with more than ten years of service, with dependents, stationed at Scott Air Force Base, making $5,799 per month base pay and $225 per month hostile fire/imminent danger pay. Housing costs were $1,638 per month, and the individual received $202.76 per month for subsistence and $250 per month in family separation allowance. The civilian profile included in the calculation was of a 46-year-old, GS-12 step 4 civilian employee from a "rest of U.S." locality area making $71,848 per year (or $34.43 per hour with an overtime rate of $35.66 per hour), working a schedule of ten-hour days/six days per week, receiving $10,000 in accidental death and dismemberment benefits, and working in an administrative career field that does not provide additional death benefits. Moreover, the pay for this civilian

Under a number of assumptions, the comparative estimate showed that deploying the profiled civilian was significantly more expensive than deploying the service member for six months ($60,016 for the civilian, compared with $48,275 for the service member). Interestingly, the *difference* in the civilian's home costs versus deployed costs was also significantly higher, as the civilian's costs when at his or her home station were lower than the service member's costs ($28,806 for the civilian, compared with $38,509 for the service member). Thus, the Air Force estimated that it cost $31,210 more to deploy a civilian than to keep him or her at the home station, whereas it only cost $9,766 more to deploy the service member (see Table 2.1).[59]

While this analysis was done several years ago, and these figures are likely outdated, they present a foundation on which to base future analysis of the comparative costs of deploying different types of personnel. However, it is critical to note that the figures in the Air Force estimate may not present a full picture of the relative costs of deploying civilians versus military personnel. Indeed, multiple considerations come into play when performing a comparative cost analysis. As indicated in the Air Force assumptions, deployed civilians are paid premiums that military members are not (e.g., overtime, hazard pay), so (assuming comparable experience) civilians may be more expensive. However, this is not the only way to think about costs. For instance, service members enjoy tax benefits that civilians do not, are granted free education and training, and earn credit toward generous military retirement pay—each of which increases the lifetime cost of military personnel. Moreover, the opportunity costs of deploying civilians instead of military personnel are difficult to capture and quantify. An example is the cost to the civilian's home office to backfill that individual while he or she is deployed (either officially in resource terms, or in terms of the extra burden placed on other office personnel to cover that position if resources to backfill the positions are unavailable).

---

was assumed not to include deductions for contributions to the retirement system and health insurance.

[59] U.S. Air Force Personnel Center, CEW Program Manager, "Military-Civilian Comparison for Six-Month Deployment to OIF/OEF," provided to RAND on July 21, 2014.

Table 2.1
Prior U.S. Air Force Calculations of Relative Costs of Military Versus Civilian Personnel at Home and for a Six-Month Deployment to OIF/OEF

	Costs ($)					
	Service Member (O-4, 10 years of service, with dependents, based at Scott Air Force Base)			Civilian (GS-12 step 4, based at Scott Air Force Base)		
Type of Compensation	Home	Deployed	Delta	Home	Deployed	Delta
Base pay/base pay and locality pay	34,794	34,794	0	35,924	35,924	0
Basic housing allowance	9,828	9,828	0	0	0	0
Basic allowance for subsistence	1,217	1,217	0	0	0	0
Per diem/meals and incidentals	0	630	630	0	630	630
Family separation allowance	0	1,500	1,500	0	0	0
Hostile fire/imminent danger pay	0	1,350	1,350	0	12,573	12,573
Foreign differential pay (35%)	0	0	0	0	12,573	12,573
Overtime pay (20 hours/ week at $35.65/hour)	0	0	0	0	18,543	18,543
Total pay (gross)	45,839	49,319	3,480	35,924	80,243	44,319
Health benefits				1,887	1,887	0
Taxable pay	34,794	0		34,037	78,356	44,319
Federal taxes	6,286	0	6,286	6,097	17,876	11,779
State income tax (Illinois = 3%)	1,044	1,044	0	1,021	2,351	1,330
Total taxes	7,330	1,044	6,286	7,118	20,227	13,109
Net pay	38,509	48,275	9,766	28,806	60,016	31,210

Table 2.1—Continued

	Costs or Coverage	
Type of Compensation	**Service Member (O-4, 10 years of service, with dependents, based at Scott Air Force Base)**	**Civilian (GS-12 step 4, based at Scott Air Force Base)**
Life insurance	$400,000 maximum coverage	$454,00 maximum benefit (includes $10,000 for accidental death)
Death gratuity	$100,000	$161,848 • $71,848 (foreign service gratuity) • $90,000 (contingency workers' compensation)
Medical coverage	100% of care	Deployed location: same as military Postdeployment: workers' compensation and/or federal employee health benefits

SOURCE: U.S. Air Force Personnel Center, CEW Program Manager, provided to RAND on July 21, 2014 (analysis conducted several years earlier).

Finally, arguments that civilians are more expensive assume that military personnel training and development is a sunk cost. However, since military training costs are extensive, this assumption is worth probing.

Taken together, the assessment of existing data on the cost of deploying civilians as opposed to the military leads us to conclude that there is a strong need for a rigorous, in-depth study of these comparative costs across all three components of the Total Force—military, civilian, and contractors.[60] Chapter Five suggests a methodology for such a study.

---

60 There has been ongoing debate regarding whether DoD civilians or contractors are more cost-efficient (in general, minus any specific consideration of deployment costs), so such a three-way cost comparison would have utility in clarifying these two workforces' relative costs as well. Then–Defense Secretary Robert Gates told the *Washington Post* in 2010, "Federal workers cost the government 25 percent less than contractors." In June 2013, Comptroller Robert Hale told a Senate subcommittee that contractors are two to three times more

Strategic Human Capital Planning Policy

A third body of policy relevant to civilian deployment—and integrally related to Total Force policy—is strategic human capital planning (SHCP) policy. An essential component of Total Force management involves strategic management of DoD's overall human capital, including the active and reserve component military, DoD civilians, and the contractor workforce. Recognizing this, Congress passed legislation in both the FY 2006 and FY 2010 National Defense Authorization Acts requiring DoD to periodically develop and submit to congressional defense committees a strategic workforce plan to shape and better manage the department's civilian workforce.[61] The intent of SHCP is described in the 2006 QDR, which states,

> To compete effectively with the civilian sector for highly-qualified personnel to build the Total Force, the Department must possess both a modern *Human Capital Strategy* and the authorities required to recruit, shape and sustain the force it needs.
>
> The new *Human Capital Strategy* focuses on developing the right mix of people and skills across the Total Force. . . . It is based on an in-depth study of the competencies U.S. forces require and the

---

expensive than civilians. At a House of Representatives hearing several months later, the Army Chief of Staff echoed Hale's remark (see J. David Cox, Sr., "DoD Insourcing Makes Financial Sense," *Federal Times*, October 29, 2013). A 2011 Center for Strategic and International Studies report argued, however, that DoD's costing methodology for comparing civilian and contractor costs does not consider a multitude of contractors' in-house costs (see David Berteau, Joachim Hofbauer, Jesse Elman, Gregory Kiley, and Guy Ben-Ari, *DoD Workforce Cost Realism Assessment*, Washington, D.C.: Center for Strategic and International Studies, May 2011). Finally, a 2013 GAO study implicitly rejected that report's findings, noting that DoD's cost estimation methodology has improved (see U.S. Government Accountability Office, *Human Capital: Opportunities Exist to Further Improve DoD's Methodology for Estimating the Costs of Its Workforces*, Washington, D.C., GAO-13-792, September 25, 2013). Again, none of these studies examined the deployment-specific costs of civilians relative to contractors.

[61] Pub. L. 111-84, National Defense Authorization Act for Fiscal Year 2010, Section 1108, October 28, 2009, codified a previous strategic workforce plan requirement that was originally enacted by Pub. L. 109-163, National Defense Authorization Act for Fiscal Year 2006, Section 1122, January 6, 2006.

> performance standards to which they must be developed. Each of the Military Departments will map the array of competencies and performance criteria that constitute its forces and also evaluate and improve personnel development processes to achieve those standards.[62]

The legislation builds on the *Interim Policy and Procedures for Strategic Manpower Planning and Development of Manpower Estimates* developed by then–Under Secretary of Defense for Personnel and Readiness David Chu in December 2003. The 2003 interim policy for strategic manpower planning states,

> Manpower authorities shall develop comprehensive, long-range forecasts of Defense manpower based on estimates of future manpower and contract support. Manpower authorities shall integrate their manpower management reporting processes, as necessary, to develop these forecasts.[63]

The policy resulting from the FY 2006 legislation, as enacted in DoDI 1400.25, states,

> It is DoD policy that a structured, competency-based approach will be instituted throughout the Department of Defense in support of strategic human capital planning (SHCP). . . . This approach will be used to identify current and future civilian workforce requirements, including those of an expeditionary nature, as part of total force planning. It will also be used to establish a plan to ensure the readiness of the civilian workforce to meet those requirements. . . . SHCP, at a minimum, will include: (1) Identifying current and projected civilian manpower requirements, including expeditionary requirements, needed to meet the Department's mission, and the strategies needed to build the civilian workforce to meet those requirements. Such

62 DoD, 2006, p. 80.

63 Chu, 2003.

> requirements should be accomplished within the context of total force planning.[64]

Meanwhile, back in 2001, GAO had placed SHCP on its list of government operations identified as high risk.[65] In a separate 2013 report, GAO stated that SHCP continued to be an area of significant concern and high risk because critical skill and competency gaps could undermine agencies' abilities to accomplish their missions.[66] Yet, several GAO reports from 2008 to 2013 nonetheless indicated shortcomings in DoD's efforts to meet its SHCP goals as delineated in policy.[67]

---

[64] U.S. Department of Defense Instruction 1400.25, *DoD Civilian Personnel Management System, Vol. 250: Civilian Strategic Human Capital Planning*, November 18, 2008, p. 2.

[65] U.S. General Accounting Office, *High-Risk Series: An Update*, Washington, D.C., GAO-01-263, January 2001.

[66] U.S. Government Accountability Office, *High-Risk Series: An Update*, Washington, D.C., GAO-13-283, February 2013.

[67] For instance, a 2008 GAO report noted that the civilian human capital strategic plan submitted by DoD in 2007 did not meet most statutory requirements, highlighting that the list of mission-critical occupations included in DoD's plan did not constitute the required assessment of skills of the existing workforce and that the plan did not address six of eight congressional reporting requirements. See U.S. Government Accountability Office, *The Department of Defense's Civilian Human Capital Strategic Plan Does Not Meet Most Statutory Requirements*, Washington, D.C., GAO-08-439R, February 6, 2008, p. 3.

A year later, GAO examined the 2008 update to the plan, noting that it showed improvement but still only partially addressed each of the FY 2006 NDAA requirements for SHCP. GAO argued that the update did not contain assessments for more than half of the 25 mission-critical occupations and that completed assessments for the remaining mission-critical occupations did not cover the required ten-year period. Furthermore, according to GAO, the plan included analyses of gaps between existing and future workforce capabilities for only approximately half of the 25 occupations and only partially addressed the requirement for a plan of action for closing such gaps. Finally, the report emphasized that while DoD's creation of a program management office responsible for monitoring updates to the plan was commendable, that office did not have and did not plan to have a performance plan articulating how the NDAA requirements would be met. See U.S. Government Accountability Office, *Human Capital: Opportunities Exist to Build on Recent Progress to Strengthen DoD's Civilian Human Capital Strategic Plan*, Washington, D.C., GAO-09-235, February 2009, p. 1.

In 2010, GAO reported that DoD's 2009 overall civilian workforce plan addressed five and partially addressed nine of the 14 legislative requirements outlined in the FY 2010 NDAA. See U.S. Government Accountability Office, *Human Capital: Further Actions*

Several elements of DoD's civilian deployment practice are relevant to overall SHCP efforts—namely, the practices guiding the recruitment, retention, and development of the DoD civilian workforce and the deployable civilian demand signal. Both are described in greater detail in the following sections.

Recruitment, Retention, and Development of the DoD Civilian Workforce

Strategic human capital planning is particularly critical with regard to expeditionary civilians, as they are more costly than their home office counterparts—at least when deployed—and because requirements for expeditionary civilian capabilities are constantly evolving and may erupt into large-scale surge requirements with little notice. Adequate planning is therefore necessary to ensure that DoD can draw on appropriate numbers of expeditionary civilians with the correct skill sets for various contingencies, on short notice when needed, either from the existing DoD workforce or from preidentified outside sources. Moreover, the challenging nature of deployments and the often unwelcome gap that deploying civilians leave in their home offices while deployed necessitate that incentives be correctly aligned to encourage expeditionary civilian deployment. Such incentives should be structured to motivate both the civilian to deploy and his or her home office to support that deployment.

Needed to Enhance DoD's Civilian Strategic Workforce Plan, Washington, D.C., GAO-10-814R, September 27, 2010, p. 5.

A 2013 GAO study noted that DoD had yet to assess the appropriate mix of its military, civilian, and contractor personnel capabilities in its strategic workforce plan as required by law. Moreover, it found that DoD had not updated its policies and procedures to reflect current statutory requirements to use its civilian strategic workforce plan and the inventory of contracted services to determine the appropriate mix of personnel to perform its mission. Finally, GAO found that DoD's policies did not reflect federal policy concerning the identification of critical functions. To remedy these shortcomings, GAO recommended that the Secretary of Defense direct OUSD(P&R) to revise DoD's existing workforce policies and procedures to address both the appropriate workforce mix and "critical functions." See U.S. Government Accountability Office, *Human Capital: Additional Steps Needed to Help Determine the Right Size and Composition of DoD's Total Workforce*, Washington, D.C., GAO-13-470, May 2013, pp. 1, 46–47.

The notion of aligning employee incentives to encourage highly qualified personnel to both join and remain in the DoD workforce is not unfamiliar. As stated in the 2006 QDR,

> The Department also needs to ensure suitable promotion and development opportunities are available to attract and retain the best and brightest military and civilian personnel. The Department's career advancement philosophy should foster innovation by encouraging career patterns that develop the unique skills needed to meet new missions such as irregular warfare.[68]

It is difficult to conclusively estimate a combination of incentives adequate to motivate appropriate numbers of DoD civilian personnel to volunteer for deployable positions. This is particularly so because, at the time of this writing, expeditionary civilian requirements were not fully integrated into planning processes, a key prerequisite for determining what constitutes "appropriate" numbers of volunteers. Yet, we do know that typical long-term, career-based incentives to encourage personnel to volunteer for any type of commitment—such as promises of career progression and increased job flexibility—were not being offered, at least not consistently. Indeed, interviews with CCMD and OSD officials indicated that under the current civilian deployment practice, civilian personnel have little career incentive to volunteer. Officials at USSOCOM noted that because there is little incentive for civilians to deploy, individuals with less experience end up deploying.[69] A senior OSD official noted a need to incentivize expeditionary civilian capabilities with the promise of future upward career mobility and flexibility, in addition to pay incentives.[70] Another senior OSD official noted that civilian deployment should be a career requirement for promotion.[71] An official at USEUCOM, meanwhile, looked at the other

---

[68] U.S. Department of Defense, 2006, p. 80.

[69] Interviews with USSOCOM officials, September 8, 2014.

[70] Interview with a senior DoD official, June 18, 2014.

[71] Interview with a senior DoD official, June 17, 2014.

side of the incentive issue, noting that the services are not incentivized to allow their civilians to deploy.[72]

Interviews with CCMD, OSD, service, and Fourth Estate officials likewise indicated that home offices have little incentive to allow or encourage their civilian personnel to deploy. A DLA official noted that the military services do not want to give up their civilians because they lose quality and continuity.[73] Along the same lines, a USEUCOM interviewee stated that the services did not like their civilians deploying because it meant losing them for a year and being required to hold their jobs.[74] An OSD official felt similarly, arguing that the Army cannot be relied upon to supply Army civilians for joint missions because it wants to keep its civilians to itself.[75]

One of the key difficulties in motivating home offices to allow or encourage their civilians to deploy is the lack of backfill available for positions vacated by civilian deployees. According to one senior USCENTCOM official, the services do not want to "give up" their civilians because they do not have the capability to backfill, and they are still paying deployees' salaries.[76] Home offices are required to keep a civilian deployee's position open for the duration of the deployment or to offer him or her a comparable position upon return. Previously, home offices were required to provide justification for the civilian's criticality to the office mission if refusing a request for that civilian to deploy.[77] However, the extent to which these requirements have held up in practice is questionable. Moreover, the requirement for a justifi-

---

[72] Interview with a USEUCOM official, September 22, 2014.

[73] Interview with a DLA official, July 29, 2014.

[74] Interview with a USEUCOM official, September 22, 2014.

[75] Interview with a DoD official, October 2, 2014.

[76] Interview with a USCENTCOM official, July 29, 2014.

[77] David S. C. Chu, Under Secretary of Defense for Personnel and Readiness, "Building Increased Civilian Deployment Capacity," memorandum, February 12, 2008; Gail H. McGinn, Deputy Under Secretary of Defense for Plans, "Department of Defense (DoD) Civilian Support to Global Expeditionary Requirements," memorandum, June 4, 2009.

cation of any refusal to allow a civilian to deploy expired in October 2011.[78]

One USSOUTHCOM interviewee noted that before the CEW was instituted with CPP at the helm overseeing sourcing, the CCMDs held service civilians "hostage," using temporary duty extensions to keep people who could not be replaced. The same interviewee argued that OSD or joint-level oversight of expeditionary civilian sourcing therefore reduces fighting between the CCMDs and the military services, adding that a cadre such as that originally envisioned for the CEW would mitigate the need for the CCMDs to "bully" the services to get civilians.[79] Other interviewees from the Navy and USCENTCOM specifically noted that a decrease in Overseas Contingency Operations funds—which covered the costs of most CEW deployments during OIF and OEF—would increase the difficulty of persuading the services to deploy civilians.[80] Finally, an OSD official noted specifically that a cadre model, such as that originally envisioned for the CEW in DoDD 1404.10, could mitigate the lack of backfill for deployed civilians, as the cadre would be preidentified, and home offices would expect to have to deploy civilians in the cadre.

The IES Office (former CEWPO) in CPP has taken steps to improve planning that may, if implemented correctly, work to realign home office incentives to deploy civilians. If home offices are better incentivized to encourage civilian deployment, their incentives may trickle down into career-based incentives for the deployees themselves. These planning improvements entail the development of a "demand signal" and a Directive Type Memorandum to mandate and direct its use, as we discuss in the next section.

Deployable Civilian Demand Signal

Specific human capital planning guidance pertaining to expeditionary civilians is included in 10 U.S.C. 2333, which requires "identification

---

78 Correspondence with a senior DoD official, June 2015.

79 Interview with a USSOUTHCOM official, September 9, 2014.

80 Interview with a U.S. Navy official, July 25, 2014; interview with a senior USCENTCOM official, August 28, 2014.

of a deployable cadre of experts, with the appropriate tools and authority, and trained in processes under paragraph (6)."[81] However, GAO noted in 2012 that this requirement had not been implemented:

> DoD has taken preliminary steps to implement its Civilian Expeditionary Workforce (CEW) policy, including establishing a program office; however, nearly 3 years after DoD's directive established the CEW, the program has not been fully developed and implemented. Specifically, DoD components have not identified and designated the number and types of positions that should constitute the CEW because guidance for making such determinations has not been provided by the Office of the Secretary of Defense. Officials stated that once key assumptions regarding the size and composition of the CEW have been finalized, implementing guidance will be issued. Until guidance that instructs the components on how to identify and designate the number and types of positions that will constitute the CEW is developed, DoD may not be able to (1) *make the CEW a significant portion of the civilian workforce as called for in DoD's fiscal year 2009 Civilian Human Capital Strategic Plan,* (2) *meet readiness goals for the CEW as required in DoD's Strategic Management Plan for fiscal years 2012–2013*, and (3) position itself to respond to future missions. [Emphasis added][82]

It is clear that management of DoD expeditionary civilians can have a significant impact on overall DoD strategic human capital management. In response to this report, the IES Office developed a "demand signal" for expeditionary civilians based on CCMD requirements from OIF, OEF, and humanitarian assistance/disaster relief missions, as delineated in eight data sources spanning the period from 2008 to 2014. Additionally, the IES Office relied on deployable civilian capability requirements to support future contingency and humanitarian assistance/disaster relief operations, as identified by USSOUTHCOM,

[81] U.S. Code, Title 10, Section 2333, Joint Policies on Requirements Definition, Contingency Program Management, and Contingency Contracting, (c)(3) and (f)(5).

[82] U.S. Government Accountability Office, *Afghanistan: Improvements Needed to Strengthen Management of U.S. Civilian Presence*, Washington, D.C., GAO-12-285, February 27, 2012.

USNORTHCOM, USAFRICOM, and USPACOM as part of a strategic review directed by the Office of the Deputy Assistant Secretary of Defense for Program Support in 2013. Such future requirements were, in turn, to be based on campaign plans and historical data from operations following the Haiti earthquake, Hurricane Katrina, and Hurricane Sandy and were to consider where DoD civilian skill sets or contracted support could augment or substitute for military personnel.[83] Using these data and a methodology developed by a working group of officials from OSD, the services, and the Fourth Estate, the deployable civilian demand signal was devised to be "the number of civilians in particular career fields who should be available to deploy in support of a broad range of expeditionary requirements." The demand signal is expressed as the percentage of civilians in the "top deployed civilian career fields" to be designated as deployable to support expeditionary requirements.[84] The top 30 occupational series included in the demand signal are shown in Table 2.2. At the time of this writing, the Defense Civilian Personnel Advisory Service and personnel from the IES Office were in the process of launching the demand signal as part of the overall effort to revise DoD policy related to expeditionary civilians.[85]

This revision of policy is intended to implement the requirement in 10 U.S.C. 2333 that DoD develop a deployable cadre of experts and DoDD 1404.10's requirement that DoD identify a "subset" of deployable civilians, as well as to respond to GAO's 2012 recommendations that DoD provide guidance on the "size and composition of the CEW."[86] More specifically, the purpose of revising this policy is to provide guidance for DoD components to identify E-E and NCE

[83] Office of the Assistant Secretary of Defense for Logistics and Materiel Readiness, "Future Combatant Command Requirements for Deployable DoD Civilians," memorandum, April 11, 2013.

[84] Office of the Deputy Assistant Secretary of Defense for Civilian Personnel Policy/Defense Civilian Personnel Advisory Service, "Deployable Civilian Demand Signal" briefing slides, April 23, 2014.

[85] Julie Blanks, International/Expeditionary Support Office, "Deployable Civilian Demand Signal," briefing, October 2014.

[86] 10 U.S.C. 2333; DoDD 1404.10, 2009; GAO, 2012.

Table 2.2
Top 30 Deployable Civilian Occupational Series

Top 30 Deploying Occupational Series (2001–2014)									
0018	**0130**	**0132**	**0201**	**0301**	**0303**	**0340**	**0343**	**0346**	**0401**
Safety and Occupational Health Management	Foreign Affairs	Intelligence	Human Resources Management	Miscellaneous Administration and Program	Miscellaneous Clerk and Assistant	Program Management	Management and Program Analysis	Logistics Management	General Natural Resources Management and Biological Sciences
0501	**0511**	**0801**	**0802**	**0809**	**0810**	**0830**	**0855**	**0856**	**1101**
Financial Administration and Program	Auditing	General Engineering	Engineering, Technical	Construction Control, Technical	Civil Engineering	Mechanical Engineering	Electronics Engineering	Electronics Technical	General Business and Industry
1102	**1104**	**1515**	**1670**	**1712**	**1811**	**1910**	**2210**	**2604**	**5803**
Contracting	Property Disposal	Operations Research	Equipment Services	Training Instruction	Criminal Investigating	Quality Assurance	Information Technology Management	Electronics Mechanic	Heavy Mobile Equipment Mechanic

SOURCE: Office of the Deputy Assistant Secretary of Defense for Civilian Personnel Policy/Defense Civilian Personnel Advisory Service, 2014.

civilians to meet the deployable civilian demand signal; to categorize E-E and NCE civilians into one of three readiness targets (immediate: 1–45 days; urgent: 46–120 days; or routine: 120+ days) pursuant to Chairman of the Joint Chiefs of Staff Instruction 1303.01E; to ensure that E-E and NCE positions are coded as such in the Defense Civilian Personnel Data System (DCPDS); to develop a strategy to fulfill each component's contribution that includes a plan for any increase in CCMD requirements for expeditionary civilians; and to annually review E-E and NCE positions to project and manage the supply of DoD civilians and ensure that the supply still fulfills the demand signal.[87]

The deployable civilian demand signal is DoD's first in-depth SHCP effort pertaining specifically to expeditionary civilians, and it is commendable in its level of depth and in the analysis supporting it. However, as we note in Chapter Five, further work remains to ensure that it is broadly applicable across potential future mission sets and that it is fully implemented as directed.

Ability to Meet Operational Needs

Because DoD expeditionary civilians are seen as "an integral part of the Total Force" and are intended to serve in "a variety of positions to provide essential capabilities" and "support mission requirements" across a wide range of contingencies, it is critical for the long-term viability of expeditionary civilian capabilities that they be seen as reliably able to meet operational needs.[88] As noted in Chapter One, the CEW's creation was driven by a confluence of events in the conflicts in Iraq and Afghanistan between 2007 and 2009.[89] Taken together, these

---

[87] Blanks, 2014.

[88] DoDD 1404.10, 2009.

[89] Interview with a former senior DoD official, June 17, 2014; interview with a DLA official, July 29, 2014; interviews with DoD officials, October 16, 2014; interview with researchers conducting a related study at National Defense University's Center for Complex Operations, June 16, 2014.

events greatly increased demand for expeditionary civilian capabilities while simultaneously hampering the effective supply of these personnel. One of the major lessons of this period was that DoD civilian deployment policies and procedures must be scoped to provide sufficient numbers of civilians to fulfill surge requirements under situations of high demand in particular contingencies. However, to better understand the potential future demand for expeditionary civilian capabilities, it is critical to develop a baseline comprehension of the various CCMDs' projected requirements for expeditionary civilian capabilities, as well as the extent and nature of CCMD thinking on how each CCMD might utilize expeditionary civilians in the future. Chapter Three explores the operational side of expeditionary civilian deployment to examine each of these issues in turn.

CHAPTER THREE

Combatant Command Utilization of DoD Expeditionary Civilians

The operational side of DoD civilian deployment comprises several elements—the assignment of requirements, sourcing, readiness preparation, and tracking—that together culminate in the utilization of expeditionary civilian capabilities by the CCMDs. While prior studies of DoD civilian deployment and the CEW have focused on the composition of the workforce, considerations for CEW personnel working alongside interagency partners, and problems with the implementation of the CEW program and other efforts to deploy expeditionary civilians, there has been no detailed analysis of the CCMDs' need for and understanding of expeditionary civilian capabilities.[1] This chapter seeks to fill that void, addressing (1) the operational processes through which requirements are assigned and through which expeditionary civilians are tasked to the CCMDs, prepared to deploy, and tracked during and after deployment; (2) CCMD requirements for expeditionary civilians in recent years; and (3) CCMD perspectives on the utilization of expeditionary civilians. In conducting this analysis, we found that the CCMDs vary widely in their projected use of expeditionary civilians and their understanding of expeditionary civilian capabilities. We also found that sourcing delays are one of the key reasons that CCMDs feel that expeditionary civilians are not a reliable labor source. When

[1] See, for instance, Dunigan, 2012; also discussed during an interview with researchers conducting a related study at National Defense University's Center for Complex Operations, June 16, 2014. As noted in Chapter One, CCMD perspectives were absent even from the CEW comprehensive review conducted by the Defense Civilian Personnel Advisory Service in late 2012.

considered in the context of the benefits and drawbacks inherent in the four different approaches, or models, used to source DoD expeditionary civilians at various times both prior to and following the release of DoDD 1404.10 in 2009, these findings highlight lessons for the future deployment of DoD civilians.

The remainder of the chapter begins with an overview and assessment of how operational processes for DoD civilian deployment have evolved since the CEW was established in 2009. The following section presents recent data on expeditionary civilian requirements from several CCMDs, followed by interview findings pertaining to CCMD utilization of expeditionary civilians. The chapter concludes with several observations from the data regarding how best to broaden CCMD utilization of DoD expeditionary civilian capabilities.

Defining the Steps in the Process

As noted earlier, the operational side of DoD civilian deployment consists of four categories of operations: assignment of requirements, sourcing, readiness preparation, and tracking.

The first step in the operational process of civilian deployment involves the formal *assignment of expeditionary civilian requirements* to the various DoD components in the Secretary of Defense Operations Book. As noted later, some expeditionary civilian requirements—such as those to fill PRTs—are most likely to fall within ad hoc and provisional organizations that the force development and management processes do not anticipate. Such expeditionary civilian positions would fall outside of the typical assignment of requirements envisioned in the GFM process and, because they would not be assigned to a force provider, would be categorized as CEW positions. While this report looks at both CEW and assigned expeditionary civilians, it is critical to understand this distinction between the two types of expeditionary civilian requirements.

The term *sourcing* has two related meanings in the context of DoD civilian deployment. In its most basic sense, sourcing refers to the provision by the force providers of able-bodied candidates for expeditionary civilian positions. However, sourcing also refers to the process of recruiting, screening, and selecting candidates for deployable positions. In the discussion that follows, we refer to sourcing in this second sense of the term. Here, sourcing comprises the first set of steps in the process of filling expeditionary civilian positions for the CCMDs.

Readiness preparation, meanwhile, comprises the second set of steps in this process. Readiness preparation primarily entails the fulfillment of necessary predeployment training and receipt of medical and security clearances, as well as anything else a deployee must do between being selected for a deployable position and actually deploying.

Finally, tracking entails the third set of steps in the operational process: keeping track of the civilian deployee while he or she is deployed, and continuing to track the deployee for some period of time postdeployment to keep him or her apprised of deployment-related announcements, to screen for deployment-related health conditions, or to notify him or her of any postdeployment awards or benefits.

The Larger Context of the Global Force Management Process

As outlined in Joint Publication 5-0, the GFM process serves three primary functions: (1) it guides the global sourcing processes for CCMD force requirements; (2) it offers the Joint Staff and force providers a decision framework for making assignment and allocation recommendations to the Secretary of Defense and apportionment recommendations to the Chairman of the Joint Chiefs of Staff;[2] and (3) it allows the Secretary of Defense to make proactive, risk-informed force management decisions.[3] The GFM process is a rather lengthy and complicated one that assigns, allocates, and apportions "large chunks of military force structure"[4] for presidential or Secretary of Defense–approved

---

[2] Per GFM guidance, there are several "force providers" for different force types. The military services are the force providers for conventional forces. USSOCOM is the force provider for special operations forces. U.S. Transportation Command (USTRANSCOM) is the force provider for mobility forces. U.S. Strategic Command (USSTRATCOM) is the force provider for strategic and intelligence, surveillance, and reconnaissance forces.

[3] U.S. Joint Chiefs of Staff, 2011b, pp. II-4.

[4] Interview with a former senior DoD official, October 2014.

military operations but is also the vehicle through which DoD civilians are authorized to deploy. To be clear, the force requirements generation process does not make specific distinctions between military and civilian personnel because it merely specifies what type of force is needed by the combatant commander. In other words, it is a process that requests and validates *capabilities.*

That being said, the allocation process, as illustrated in Figure 3.1, begins with the combatant commander identifying requirements.

These requirements can originate through rotational force requirements identified in operational plans, operational plans in concept format, or emerging force requirements (i.e., requirements that cannot be met by available assigned forces or forces already allocated).[5] The combatant commander validates each requirement, which is then assigned a Force Tracking Number. This number is then forwarded electronically to the Joint Staff through an official RFF message. The Joint Staff then validates and prioritizes each request based on the Secretary of Defense's *Guidance for Employment of the Force* and other sourcing guidance and assigns a force provider to supply a sourcing solution. The force providers generate a sourcing solution, draft modifications to their respective GFM Allocation Plan (GFMAP) annexes, and forward recommendations to the Joint Staff. The Joint Staff then forwards GFMAP modifications to the CCMDs, services, combat support agencies, other DoD agencies, and OSD and briefs the solution (through the Chairman of the Joint Chiefs of Staff and the GFM Board) to the Secretary of Defense, who ultimately makes the decision.

---

[5] Note that the definitions for assigned and allocated forces, per Title 10, are as follows:

> Assignment: The President, through the Unified Command Plan (UCP) documents his direction for assigning forces for Unified Commands. (10 U.S.C. 161, 162, and 167)
>
> Allocation: By the authority of the SecDef, forces assigned to a combatant command may be transferred or allocated to another combatant commander for employment . . . under procedures prescribed by the SecDef and approved by the President. (10 U.S.C. 162)

Additionally, the Chairman of the Joint Chiefs of Staff can "apportion" forces:

> Apportionment: The [Chairman] is responsible for preparing strategic plans including projected resource levels . . . [and] apportions forces to combatant commands based on the SecDef's contingency planning guidance. (10 U.S.C. 153)

Figure 3.1
Global Force Management Allocation Process

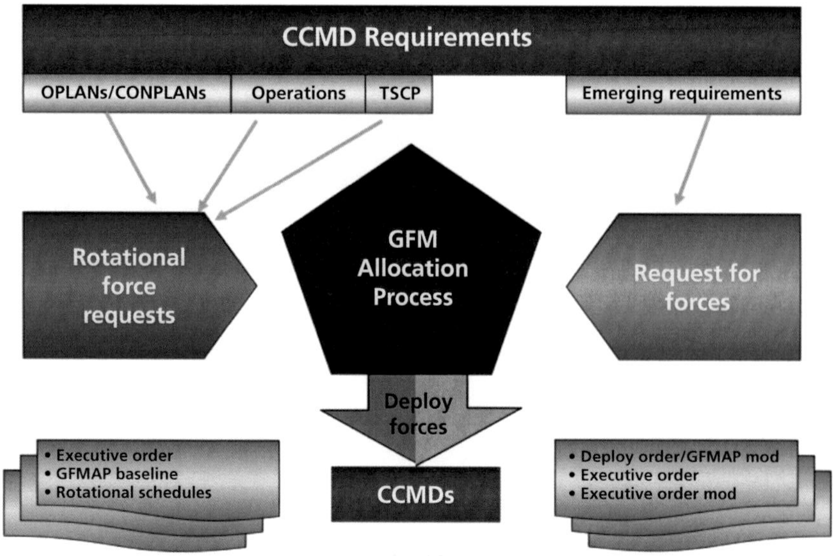

SOURCE: Adapted from Joint Forces Staff College, "Introduction to Global Force Management," briefing on the Joint Operations Planning Process, March 3, 2012, slide 13.

NOTES: OPLAN = operation plan. CONPLAN = concept of operations plan. TSCP = theater security cooperation plan.

RAND *RR975-3.1*

After approval by the Secretary of Defense, the Joint Staff publishes the modifications to the GFMAP annexes.[6]

DoD civilians can be deployed to fill force requirements identified in any of the GFMAP's four annexes because the GFMAP gives force providers flexibility in identifying the appropriate unit for deployment.[7] While most sourcing solutions are considered "standard force

---

[6] U.S. Joint Chiefs of Staff, 2011b, p. H-2. See also Patrick Sweeney, *A Primer for: Guidance for Employment of the Force (GEF), Joint Strategic Capabilities Plan (JSCP), the Adaptive Planning and Execution (APEX) System, and Global Force Management (GFM)*, U.S. Naval War College Joint Military Operations Department, July 2013, pp. 16–20.

[7] Annex A is for conventional forces, with the services acting as the force providers; Annex B is for special operations forces, with USSOCOM acting as the force provider;

solutions," nonstandard sourcing categories, such as "in-lieu-of" and "ad hoc," allow for DoD civilian sourcing solutions when preferred military sources are not available or in situations in which civilians are required, such as was the case for staffing PRTs.[8] DoD civilians are also explicitly identified in GFMAP Annex D, which specifies JIA forces, or temporary manpower requirements to augment a joint task force (JTF) headquarters during a contingency. The JIA sourcing allocation process mirrors the RFF sourcing process except that the format and information requirements of the JIA request are electronically submitted through the Electronic Joint Manpower and Personnel System (eJMAPS) and are not assigned Force Tracking Numbers. After validation by the Joint Staff, these JIA requirements are codified in CCMD joint manning documents (JMDs; see Figure 3.2).[9]

To summarize, the GFM process provides military capabilities to combatant commanders, but the allocation processes—specifically, the RFF and JIA allocation functions—are the vehicles through which DoD civilians are identified as sourcing solutions to meet commanders' needs. However, it should be noted that although the GFM process allows DoD civilians to be identified as sourcing solutions, there is considerable debate about the degree to which DoD civilians are considered or included in the requirements generation process. As mentioned earlier, some expeditionary civilian requirements—those of ad hoc or provisional organizations not anticipated in the GFM process—are not visible to GFM planners, and are thus not assigned to force providers during this process. It is these expeditionary civilian requirements that have, in the past, tended to be categorized as CEW requirements. One potential remedy to this situation would be for operational planners to identify the need for CEW capabilities and individuals in operational plans and to ensure that the planning scenarios for sizing and shaping the force address similar requirements.

Annex C is for mobility forces, with USTRANSCOM acting as the force provider; and Annex D is for JIA, with the services acting as the force providers. See U.S. Joint Chiefs of Staff, 2011b, p. H-1.

8 U.S. Joint Chiefs of Staff, 2011b, pp. H-3–H-4.

9 U.S. Joint Chiefs of Staff, 2011b, pp. H-2–H-4.

Figure 3.2
Joint Individual Augmentation Process Flow Diagram

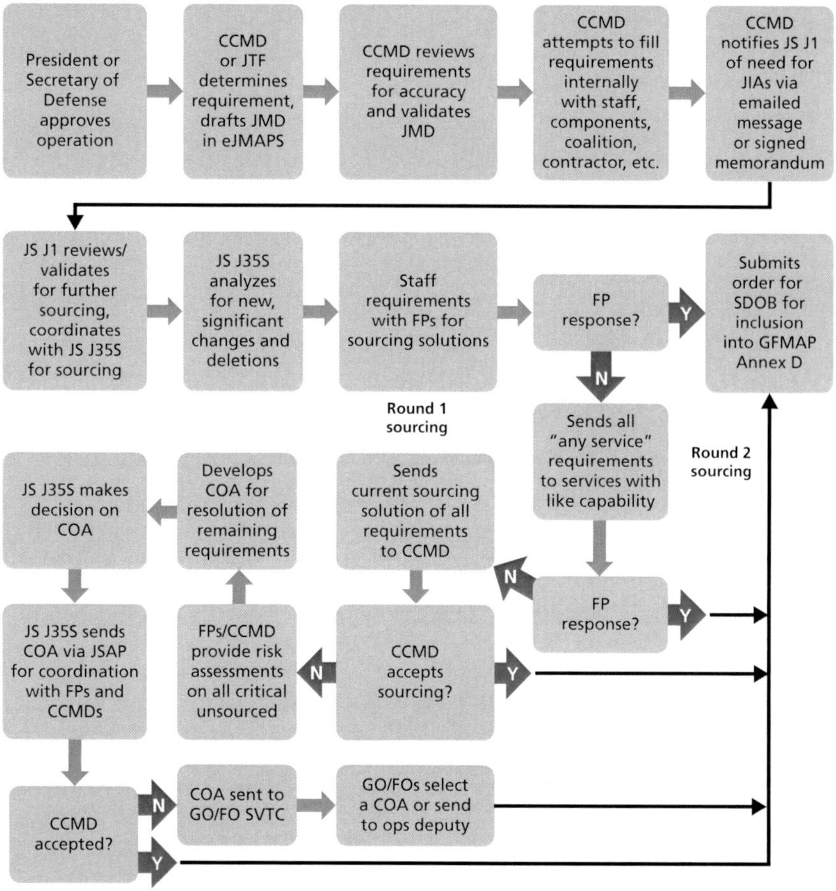

SOURCE: Adapted from Chairman of the Joint Chiefs of Staff Instruction 1301.01F, *Joint Individual Augmentation Procedures*, November 17, 2014, Enclosure B.

NOTES: JS = Joint Staff. J35S = joint force coordinator office. FP = force provider. SDOB = Secretary of Defense Operations Book. COA = course of action. JSAP = Joint Staff Action Package. SVTC = secure video-teleconference. GO/FO = general/flag officer.

RAND *RR975-3.2*

Different Operational Models for DoD Civilian Deployment

DoD has experimented with various methods and operational models for the overall operational process pertaining to civilian deployment, each with its sourcing aspects tied to the GFM process.

CCMD Executive Agent Model

Prior to the 2009 establishment of the CEW, sourcing, readiness, and tracking were all managed in a decentralized fashion by the military services, DoD Fourth Estate agencies, and the CCMDs' executive agents.[10] In other words, a CCMD would send a requirement for an expeditionary civilian to its executive agent, which would then look for a civilian to fill it from among its service's own ranks. Once an individual was selected, the DoD component that "owned" the civilian in question was responsible for providing any necessary predeployment training and ensuring that medical and security clearances were obtained. Moreover, each of the individual DoD components was responsible for tracking its deployed civilians during and after deployment.

This process aligned with the service responsibilities for force provision outlined in Title 10 but had several disadvantages. For example, the services were not motivated to provide their best personnel

---

[10] A DoD executive agent is defined as follows: "The Head of a DoD Component to whom the Secretary of Defense or the Deputy Secretary of Defense has assigned specific responsibilities, functions, and authorities to provide defined levels of support for operational missions, or administrative or other designated activities that involve two or more of the DoD Components" (U.S. Department of Defense Directive 5101.1, *DoD Executive Agent*, September 3, 2002). With regard to executive agency for the CCMDs, DoD policy states, "The Secretaries of the Military Departments shall provide or arrange for the administrative and logistic support of the headquarters of the Combatant Commands and the U.S. Element, North American Air Defense Command."

Support of each CCMD is designated to the services as follows: USEUCOM: Secretary of the Army; USSOUTHCOM: Secretary of the Army; U.S. Joint Forces Command: Secretary of the Navy; USPACOM (except U.S. Forces Korea): Secretary of the Navy; U.S. Forces Korea: Secretary of the Army; USCENTCOM: Secretary of the Air Force; USSTRATCOM: Secretary of the Air Force; U.S. Element, North American Aerospace Defense Command: Secretary of the Air Force; USSOCOM (except Joint Special Operations Command): Secretary of the Air Force; Joint Special Operations Command: Secretary of the Army; USTRANSCOM: Secretary of the Air Force; and USNORTHCOM: Secretary of the Air Force. See U.S. Department of Defense Directive 5100.3, *Support of the Headquarters of Combatant and Subordinate Joint Commands*, certified current as of March 2004.

for deployable positions because they did not want to lose them from their home offices without backfill, and the relevant executive agent did not have visibility over the entire pool of potential civilian candidates, nor the capability to hire across the other military services in a joint manner. In the words of one CCMD official, "The services are not incentivized to let their civilians go."[11]

As a result, when demand for expeditionary civilian capabilities surged in 2007–2009 (as discussed in Chapter One), force providers operating under this model were unable to supply adequate numbers of personnel to meet this demand.[12] Finally, under such a stovepiped model of operations focused specifically on the workforce of the executive agent, the level and quality of civilian tracking varied depending on the executive agent in question, translating into differential treatment of different members of the workforce based on which service or entity "owned" them.[13]

Another problem with the executive agent model was that it proved unable to provide civilians on a rotational basis. This lack of timely replacements, or rotations of civilians, meant that many civilians' deployments were extended. When this happened, according to one CCMD official, the service home offices "fought the CCMDs to get their people back." This interviewee referred to this as CCMDs holding service civilians "hostage."[14] Another OSD official noted, "You can't count on the Army to supply their civilians for joint missions. They want to keep them to themselves."[15] In sum, said a former senior DoD official, "The executive agent model breaks down if it's service-oriented."[16]

---

[11] Interview with a USEUCOM official, September 22, 2014.

[12] Interview with a former senior DoD official, June 17, 2014; interview with a DLA official, July 29, 2014.

[13] Interviews with DoD officials, October 16, 2014.

[14] Interview with a CCMD official, September 9, 2014.

[15] Interview with a DoD official, October 2, 2014.

[16] Interview with a former senior DoD official, June 17, 2014.

CEWPO Model

With the establishment of the CEW, this process shifted dramatically. DoDD 1404.10 placed CPP—traditionally a policy office—in charge of both the policy and operational aspects of DoD civilian deployment.[17] This meant, in effect, that CPP, and the CEWPO specifically, became a force provider for the Joint Staff and would coordinate across the DoD components to find candidates to fill requirements not assigned to a specific component or force provider. Under this model, the CEWPO collected and screened resumes, attempting to keep a database of a large number of potential candidates (at one point numbering over 17,000) up to date, and liaising with the services and other force providers across the joint DoD civilian workforce to identify viable candidates.[18]

The CEWPO often used word of mouth to get out position announcements; if a capability did not exist or appear to be available within DoD, it would look outside of DoD. Significantly, during this period, the CEWPO made extensive use of Schedule A hiring authorities to bring in temporary employees from outside the existing DoD civilian workforce. Indeed, approximately 50 percent of the civilians deployed under the CEW program when managed by the CEWPO were hired under Schedule A authorities.[19] The CEWPO was also responsible for coordinating deployees' readiness preparation under this model. In doing so, it worked with the Office of the Deputy Assistant Secretary of Defense for Readiness to send selected candidates to Camp Atterbury for predeployment training and to coordinate medical and security clearances.[20]

As with the original operational model for civilian deployment, there were both advantages and disadvantages. One of the model's ben-

---

[17] DoDD 1404.10, 2009, Enclosure 2, pp. 11–12.

[18] Dunigan, 2012; interviews with DoD officials, May 30, 2014.

[19] Interviews with DoD officials, May 30, 2014. This is also documented in a CEWPO briefing, which states that in May 2013, 216 out of 570 (38 percent) "boots-on-the-ground" CEW deployees were Schedule A hires. In December 2012, it was 252 out of 477 (53 percent). And in October 2012, it was 251 of 401 (63 percent). See Blanks, 2013, slides 14–15.

[20] Interviews with senior DoD officials, October 2, 2014.

efits was that CPP and the CEWPO understood the unique nature of DoD civilian employment and the administrative requirements of civilian hiring, which fundamentally differ from military hiring requirements and processes. Another benefit was that, at least in theory, the CEWPO had joint oversight and visibility over the entire potential candidate pool. The ability to utilize Schedule A authorities was beneficial in that surge requirements during this period caused demand for expeditionary civilian capabilities to outstrip supply; it also meant that the CEWPO had the ability to hire people not already "owned" by other force providers, which increased the ease of hiring because it did not entail fighting with the force providers for their existing employees.

In practice, however, the CEWPO was not allocated any resources to take on the operational responsibilities of the CEW program. This meant that the office was understaffed and the potential benefits of this operational model were unrealized to some extent. Several other disadvantages were associated with this model. First, the CEWPO and CPP did not "own" any of the candidates that they were trying to recruit—causing the CEWPO to feel, in the words of one former official, as if "they were begging, borrowing, and stealing" assets to deploy.[21] Second, reliance on temporary hiring authorities is useful only in a surge situation; for steady-state requirements, temporary hiring authorities are disadvantageous in that they hinder the development of a standing workforce with the capabilities and skill sets needed over the long term. Third, delays under this model were notable and were sufficiently long—at times, 120 days or more to deploy an identified individual—to lead several CCMDs to develop a lasting distaste for the utilization of expeditionary civilian capabilities.[22] Finally, at least one CCMD official noted that the CEWPO was "difficult to deal with" and "it was tough to get any information from them."[23] This was likely due to the CEWPO's resource and staffing constraints, but it had downstream ramifications that ultimately shaped at least one CCMD's perceptions of the utility of the CEW program in general.

---

[21] Interview with a former DoD official, April 30, 2014.

[22] Interviews with CCMD officials, September 30 and November 5, 2014.

[23] Interview with a CCMD official, November 5, 2014.

It is critical to note that tracking of civilian deployees during and after deployment appears to have been woefully inadequate under this model. This was because the CEWPO never took on this responsibility, leaving it to the home offices in keeping with the policy laid out in DoDD 1404.10, and because the home offices were not incentivized to look favorably upon civilians deploying out of their offices. National Defense University researchers who studied the history of the CEW told us, "Nearly one-third of our interviewees lost their jobs while deployed. Many more felt disgusted and pained by the response of their home command to their deployment."[24] Notably, CPP officials disputed the accuracy of this claim, pointing out the "virtual impossibility" of a CEW deployee completely losing his or her position as a result of deployment. The more likely scenario, according to these officials, is that these individuals were assigned to a different position from their predeployment position upon their return, or they were Schedule A employees who, by definition, were temporary employees hired only for the term of their deployment.

Nonetheless, the tracking problem does appear to have had real ramifications for deployees. Stories abound of home office supervisors being responsible for writing performance reviews for CEW deployees during the period of their deployment, even though they did not have visibility over the deployee's activities and therefore could not adequately comment upon performance. One former senior OSD official noted, "Rating reports were often derogatory because supervisors didn't understand."[25] Another Joint Staff official who had deployed as a member of the CEW recounted that a mistake was made in his pay for the last stage of his deployment—one that took almost a year to remedy because there was no office to champion or follow through on the deployee's concerns.[26]

Moreover, although DoDD 1404.10 outlines a requirement for both a deployment index and a readiness index, neither was created

---

[24] Interview with researchers conducting a related study at National Defense University's Center for Complex Operations, June 16, 2014.

[25] Interview with a former DoD official, June 17, 2014.

[26] Interview with a senior DoD official, June 17, 2014.

until the IES Office developed the deployable civilian demand signal, as discussed in Chapter Two. As a result, under CEWPO management of CEW operations, there was no standardized method in place to systematically track the readiness of CEW personnel at the OSD level.[27] Readiness of service-owned civilians is supposed to be tracked in the Deployment Readiness Reporting System (DRRS), a military system accessible to the CCMDs for readiness tracking. This system attempts to measure readiness at the strategic level, gathered from numerous metrics at individual units, to ensure that the military as a whole can meet the objectives laid out in the National Military Strategy.[28] CPP officials noted that it is more cost-effective for the services to track civilian readiness through DRRS than for CPP to track it centrally in DCPDS.[29] Yet, such a view does not consider how to track expeditionary civilians owned by Fourth Estate agencies or externally hired temporary employees, as these personnel would not fall into the individual service's tracking schemes.[30]

---

[27] In fact, CPP tried to pass operational responsibilities for the CEW program to the Office of the Deputy Assistant Secretary of Defense for Readiness at least once during this period, but because the responsibility did not come with resources to support it, that office declined to take it on. Interviews with DoD officials, April 30 and October 16, 2014.

[28] DRRS was designed to replace the Status of Resources and Training System, an older DoD reporting system that was more narrowly constructed and had a limited ability to measure unit readiness.

[29] Interviews with DoD officials, April 30, 2014. There is no crosswalk between DRRS and DCPDS, the IT system used to code U.S. federal government civilians by their positions. Interviewees thought that updating DCPDS to track available civilian personnel by skill set, or crosswalking it with DRRS to track civilian readiness for deployment, would likely entail a wholesale revision of the DCPDS software, in addition to a directive to the military services to enter the coded data and a directive to civilian personnel to continually update their own skill sets and other relevant information. Interviews with DoD officials, October 16, 2014. It is important to note that these were interviewee perceptions, and it was not within the scope of our study to verify whether updating DCPDS or crosswalking it with DRRS in this way would actually entail a wholesale revision of the DCPDS software.

[30] The IES Office is looking to Washington Headquarters Services to be the central point of contact for all Fourth Estate civilians who deploy. However, the extent to which Washington Headquarters Services will track or oversee Fourth Estate expeditionary civilians remains unclear.

On the whole, it appears that the "joint" model of operational oversight led by the CEWPO might have worked effectively if it had been adequately resourced and structured so that force providers were incentivized to provide their civilian personnel without complaint, and if tracking had been standardized across the various force providers and managed centrally.

Army G-1 Model

The challenges that the CEWPO faced in managing the operational side of the CEW program led to a decision to turn over all CEW operational responsibilities to the Army G-1 (Personnel) in March 2014.[31] The Army G-1 managed the program much as the CEWPO had in that it relied largely on word of mouth and a website to advertise the CEW and available positions, as well as to urge interested candidates to apply.[32] The G-1 maintained a database of resumes, but unlike the CEWPO it kept resumes for a maximum of six months. According to Army officials, the G-1 also tried to close gaps in the CEW sourcing process—for instance, by identifying qualifications for deployment and vetting candidates more carefully at the outset. As part of this vetting process, the G-1 adopted a process started by the Air Force's CEW liaison office, and eventually also followed by the CEWPO, of accepting resumes only from candidates who had received preapproval to deploy from their home offices. Neither the Navy nor DLA followed such a preapproval process, but according to Air Force and Army officials, this process has been shown to greatly reduce sourcing delays.[33] For the Air Force, adopting such a process change allowed its CEW liaison office to reduce deployment wait times from 120 days to 45 days from the time of selection to the time of deployment.[34] The Army G-1 reported that it was able to reduce the predeployment process from 120+ days to

---

[31] Interviews with DoD officials, May 30, 2014.

[32] Interviews with DoD officials, May 30, 2014; interviews with U.S. Army officials, July 18, 2014.

[33] Interviews with U.S. Army officials, July 18, 2014; interview with an Air Force official, July 21, 2014.

[34] Interview with an Air Force official, July 21, 2014.

between 45 and 60 days. It should be noted that CPP officials dispute this, arguing that the CEWPO had reduced the predeployment process time frame by early 2014 to approximately 45–60 days, indicating that this was the existing deployment time frame when the Army G-1 assumed operations. Army G-1 officials reported that shorter deployment times were achieved when E-E personnel were predesignated within the home office in question. This was because E-E readiness was regularly maintained (particularly in terms of medical clearances and acquisition of passports, which could cause the longest delays) and because home offices had planned and programmed for losses of their E-E–designated personnel and were thus more willing to let them go.[35] Notably, however, requirements assigned to E-E personnel were service-specific rather than CEW requirements.

Like the CEWPO operational model, the Army G-1 operational model has both benefits and drawbacks. The advantages of this model were similar to those in the later years of the CEWPO's existence, including a reduction in deployment times and assurance that home office approval for deployment was obtained before a candidate could apply. However, because the Army G-1 represents only one of the military services, this model ran the risk of bias in favor of hiring Army civilians for deployable positions, as well as a lack of familiarity with enterprise-wide expeditionary civilian capabilities. This is problematic because it may unintentionally and unnecessarily shrink the pool of available candidates, or at least the pool of those who are regularly considered for expeditionary positions. Indeed, CCMD officials indicated that they thought the recruitment process managed by the Army G-1 resulted in a shallow pool of potential candidates due to the lack of joint visibility. According to one CCMD interviewee, the effectiveness of the G-1 sourcing process varied on a case-by-case basis and depended on whether the CCMD in question had a specific individual in mind for a particular position. When it did, the Army G-1 process was successful; however, when it did not, there "was a lot of frustration regarding the lack of available candidates."[36] On the whole, the

35 Interviews with U.S. Army officials, July 18, 2014.

36 Interviews with CCMD officials, September 30, 2014.

interviewee reported, "The CEW program wasn't meeting our needs because there were holes in the screening process of the candidates," and even went so far as to say "[Our command] was willing to go understrength as opposed to dealing with the CEW program again."[37]

USCENTCOM CEW Office Model

USCENTCOM stood up its CEW office to manage the operational side of civilian deployment to the USCENTCOM area of responsibility (AOR) in mid-2014. The USCENTCOM CEW Office hiring process differs from that practiced by either the CEWPO or the Army G-1 in that it posts openings on the USAJobs website, thereby aiming to increase the potential applicant pool substantially. Whereas the CEWPO and Army G-1 would initially screen resumes and then send them forward to the appropriate CCMD for approval, the USCENTCOM CEW Office receives resumes directly through its USAJobs postings and screens them using in-house subject-matter experts to find "fully qualified individuals." As with the Army G-1 and Air Force processes, USCENTCOM requires that candidates receive approval to deploy from their home offices prior to submitting a position application. USCENTCOM also requires the individual to deploy if selected; he or she is not simply placed in a pool and given the option to deploy. If selected, he or she must deploy for the position or be rejected. Although USCENTCOM was in the process of standing up its CEW Office when we interviewed officials there, the aim was to be able to deploy people within 30–60 days after initial selection of the candidate.[38]

Under this model, readiness preparation—including any necessary training and medical clearances—will be managed by the services that own the civilians. According to USCENTCOM, the services are accustomed to managing these activities and have the infrastructure in place to ensure adequate readiness preparation. However, the services will have to request Overseas Contingency Operations funding to cover the added expense of preparing expeditionary civilians for

---

[37] Interviews with CCMD officials, September 30, 2014.

[38] Interview with a senior USCENTCOM official, August 28, 2014.

deployment to USCENTCOM. The model therefore depends on continued Overseas Contingency Operations funding.[39]

The establishment of the USCENTCOM CEW Office is somewhat contentious among stakeholders, including in some of USCENTCOM's own directorates; it is a completely different model and holds the potential to prioritize different interests than were prioritized by either the CEWPO or the Army G-1 in sourcing expeditionary civilian positions. Officials at USCENTCOM believe that having a CCMD CEW Office is a good model because, to date, there have been no central mechanisms to compel the services to source expeditionary civilian positions. One USCENTCOM interviewee believed that it would be easier for each of the CCMDs to have a CEW Office, because it would serve as a single point of contact or "clearinghouse" for that CCMD's requirements. The services, in contrast to this model, do not have visibility across the entire joint capability pool, resulting in the CCMDs "having to guess which service to send a requirement to."[40] It should be noted that two high-level CCMD officials reported this problem despite the fact that expeditionary civilian requirements are supposed to be sent through the Joint Staff for disbursement to the force providers. According to this logic, the CCMDs know their requirements better than anyone else and can thus screen resumes more effectively than any outside entity.

Another USCENTCOM interviewee reiterated these concerns regarding stovepiping of the capability along service lines, noting,

> If it is determined that . . . a central office will exist for managing the recruitment and staffing, I believe it is important that Theater have a single [point of contact] in [the] Continental United States (CONUS) to manage these processes. Currently these processes are managed by different agencies based on the program the civilian will deploy under (i.e., [Ministry of Defense Advisors] vs CEW). Currently this is problematic for multiple reasons to include:

---

[39] Interview with a senior USCENTCOM official, August 28, 2014.

[40] Interview with a senior USCENTCOM official, August 28, 2014.

- Theater is required to communicate requirement information to various CONUS based offices vs. a single point of contact increasing the likelihood of miscommunication.
- Each office has different business rules/processes for advertising, selecting, deploying and extending civilians, which is confusing to the Services.
- The agencies are competing for applicants from the same employee pool vs. looking at the DoD employee pool for a holistic sourcing solution.
- Separating the requirements between managing offices prohibits Theater from cross-level[ing] selectees to fill higher priority vacancies across program borders.[41]

However, there are perceived drawbacks associated with the CCMD-directed sourcing model as well. First, only USCENTCOM and USAFRICOM indicated an interest in having a CCMD CEW office, and USAFRICOM noted that it did not have the resources to support such an office.[42] Officials from other CCMDs did not foresee a sufficient need for steady-state expeditionary civilian capabilities to warrant the establishment of a resident CEW office, nor were there resources to support such an office.[43] Second, one OSD official with whom we spoke indicated that she perceived the USCENTCOM CEW office model as not well accepted in theater: "CENTCOM staff are military for the most part and do not understand the way civilian hiring/career management works. The folks in Afghanistan would like one centralized entity that understands civilian hiring to oversee expeditionary civilian sourcing and operations."[44]

---

[41] Correspondence with a USCENTCOM official, September 3, 2014.

[42] Interview with a senior USCENTCOM official, August 28, 2014; interviews with USAFRICOM officials, September 30, 2014.

[43] Interviews with USSOCOM officials, September 8, 2014; interview with a USSOUTHCOM official, September 9, 2014; interview with a USEUCOM official, September 22, 2014; interview with a USPACOM official, September 22, 2014; interviews with USSOCOM officials, September 25, 2014; interview with a USNORTHCOM official, November 5, 2014.

[44] Correspondence with an OSD official, October 29, 2014.

Judging the adequacy of any one of these operational models for future DoD civilian deployment—and assessing whether some combination of them or a different operational model altogether might be more effective—requires understanding the CCMDs' requirements for expeditionary civilians, as well as their differing perspectives on the use of expeditionary civilians. We address each of these topics in the sections that follow.

Combatant Commands' Varied Utilization of Expeditionary Civilians

CCMD Requirements for and Perspectives on the Utilization of Expeditionary Civilians

Since the creation of the CEW in 2009, USCENTCOM has been, by far, the largest CCMD customer of expeditionary civilian capabilities. In fact, Joint Staff requirements data reveal that, as of July 2014, USCENTCOM expeditionary civilian requirements constituted 89 percent of the *global* total expeditionary civilian requirement.[45] This is unsurprising, given the dual wars in Iraq and Afghanistan in the USCENTCOM AOR during this period. However, this speaks to a large degree of uncertainty regarding future requirements for expeditionary civilians across different mission sets from those that have been prominent in OIF and OEF. This is particularly so because the USCENTCOM AOR represents a relatively small region of the globe in comparison to that represented by the other geographic CCMDs (see Figure 3.3). Whether USCENTCOM will continue to be a central focus of U.S. military activity in the future remains unclear, but continuing conflict in the region may perpetuate or generate the need for expeditionary civilians.

The most pressing question is whether unforeseen demands will arise in other areas in addition to those in the USCENTCOM AOR. Mission-specific, scenario-based forecasting, discussed in terms of

---

[45] Interview with a Joint Staff official, July 25, 2014, and supporting documentation provided by the interviewee.

Figure 3.3
USCENTCOM Area of Responsibility Relative to Those of Other Geographic Combatant Commands

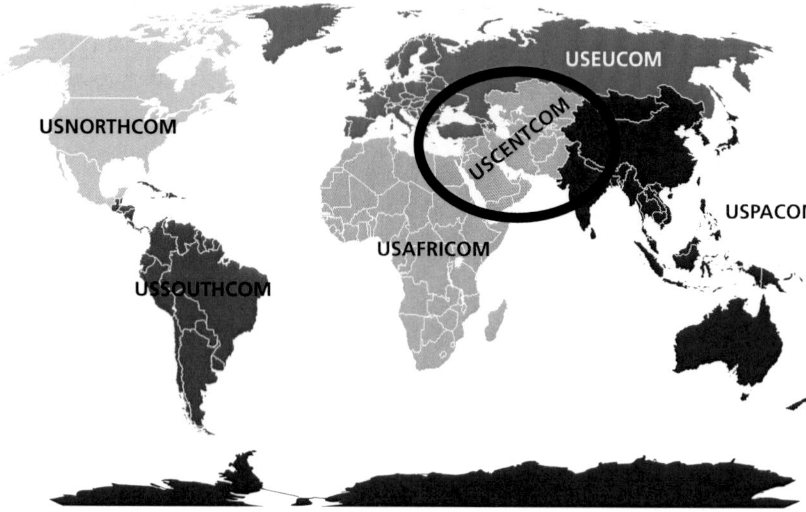

RAND *RR975-3.3*

SHCP in Chapter Two, could be very helpful in scoping out these potential future demands for expeditionary civilians in areas where they are not currently being utilized. This elaborate on this idea in Chapter Five, but it is important to note here that, for the most part, CCMDs other than USCENTCOM are not planning to use expeditionary civilians in the future, nor do they appear to appreciate the potential benefits of doing so.

Indeed, beyond USCENTCOM, CCMD requirements for expeditionary civilians are largely undefined but do vary from command to command. In response to a 2013 request for information that CPP issued to the CCMDs, USPACOM stated,

> USPACOM does not have a Civilian Expeditionary Workforce in place. [Headquarters] identifies and designates Key and Emergency Essential (E-E) personnel based on position encumbered. USPACOM Sub-Unified Commands ([U.S. Forces Japan/U.S. Forces Korea]) do not deploy and remain in place to support con-

> tingency and/or humanitarian disasters overseas. Command recognizes further guidance is needed from higher levels to code positions appropriately to support contingency operations.[46]

Similarly, USNORTHCOM stated in response to the same request that it "has no specific requirements for deployable DoD civilians," though it did note that, "where acceptable, NORTHCOM would accept a civilian in lieu of military capacity."[47] USAFRICOM and USSOUTHCOM responded to the request with small CEW requirements, but not to the same extent as USCENTCOM. For example, USAFRICOM identified more than 120 position requirements ranging from logistics planners, intelligence planners, and cyber planners to administrative personnel.[48] USSOUTHCOM, on the other hand, noted only approximately 50 requirements for expeditionary civilians, including logistics specialists, cyberspace operations officers, engineers, and administrative specialists.

These data are supported by our interview findings, which indicate that the CCMDs vary quite drastically in their actual utilization of and overall perspectives regarding expeditionary civilians. We found that the CCMDs tended to fall into one of three categories: (1) CCMDs that have no use for expeditionary civilians, (2) CCMDs that have reservations about using expeditionary civilians, and (3) CCMDs that widely use expeditionary civilians.

CCMDs That Do Not Use Expeditionary Civilians

Several CCMDs argued that they cannot utilize expeditionary civilians. This may be due to the particular mission sets of these commands, their existing in-house civilian capabilities, or their lack of understanding of expeditionary civilian capabilities, as discussed later. One USEUCOM representative we interviewed noted that USEUCOM has not deployed civilians and has no expeditionary civil-

---

46 "J1 Spreadsheets for CCMD Requirements for Deployable DoD Civilians," undated, provided to RAND by CEWPO in April 2014.

47 "J1 Spreadsheets for CCMD Requirements for Deployable DoD Civilians," undated.

48 "J1 Spreadsheets for CCMD Requirements for Deployable DoD Civilians," undated.

ians on its JMD. The interviewee specifically stated, "EUCOM has no need for expeditionary civilians or a CEW office."[49] USPACOM similarly stated that it has not deployed civilians, has no expeditionary civilian positions on its JMD, and has no need for a resident CEW office within its command.[50]

CCMDs That Are Reluctant to Use Expeditionary Civilians

Some CCMDs argued that using expeditionary civilians is either administratively burdensome or takes too long to be operationally useful. These CCMDs could feasibly have a need for expeditionary civilian capabilities but are hesitant to utilize them. For example, USNORTHCOM reported that it has no need for expeditionary civilians because it primarily operates domestically within the continental United States. However, USNORTHCOM representatives reported that they do "deploy" their headquarters civilian staff to locations across the continental United States using temporary duty orders. Because USNORTHCOM deploys these individuals on a very quick-turn basis (i.e., within a week or two), they noted that the ability to utilize DoD expeditionary civilians for these tasks would be attractive only if the system were changed so that expeditionary civilians could deploy within a week or two of being selected for a position.[51]

USAFRICOM, meanwhile, noted that it has used expeditionary civilians in the past (as indicated by the 2013 requirements data referenced earlier) but began deploying its own headquarters civilian staff because the CEW program was not meeting the command's needs due to "holes in the screening process" for candidates.[52] As noted earlier,

---

[49] Interview with a USEUCOM official, September 22, 2014.

[50] Interview with a USPACOM official, September 22, 2014.

[51] Interview with a USNORTHCOM official, November 5, 2014.

[52] Notably, the USAFRICOM representatives with whom we spoke indicated that their difficulties in dealing with the CEW program were likely due to the unique skill sets needed both to deploy to Africa and to handle the issues that this "baby command" faced at the time, just as it was being established (interviews with USAFRICOM officials, September 30, 2014). Yet, a civilian deployment concept needs to be prepared to respond to a variety of unforeseen requirements from a variety of theaters to be flexible enough to be

USAFRICOM went so far as to say it was "willing to go understrength as opposed to dealing with the CEW program again."[53]

USSOUTHCOM also marginally falls into this category: While it does utilize expeditionary civilians to some extent, an interviewee noted that it uses them only as a "last resort" because it is "a hassle to get them here, get them ready."[54] He attributed the burdensome nature of deploying DoD civilians for USSOUTHCOM missions to the difficulty in compelling civilians to work overtime on deployment and in commanding them to complete various deployment tasks in general. "Civilians have different rights that military personnel do not have; they have the mindset that they don't have to do everything the military asks of them."[55]

CCMDs That Widely Use Expeditionary Civilians

As noted previously, USCENTCOM has been the predominant user of expeditionary civilians, and, according to a USCENTCOM official, the command has consistently employed between 400 and 600 expeditionary civilians per year since 2009.[56] In 2009, approximately 70 percent of these civilians were used to source JIA requirements, whereas only 30 percent were used to fill holes in RFFs; however, by 2014, the percentages for JIA and RFF expeditionary civilians had equalized.[57] Nevertheless, according to various interviewees, these expeditionary civilians were used largely to circumvent force management levels and not to fill low-density/high-demand requirements for particular civilian skill sets. This fairly extensive use of DoD expeditionary civilians, and the practice of using them simply as additional boots on the ground, constitutes the third category of CCMD utilization of deployed civilians. Again, it is notable that this category

---

viable over the long term. Therefore, USAFRICOM's experience is quite instructive for DoD as it looks beyond OIF and OEF and scopes out a long-range plan for civilian deployment.

[53] Interviews with USAFRICOM officials, September 30, 2014.

[54] Interview with a USSOUTHCOM official, September 9, 2014.

[55] Interview with a USSOUTHCOM official, September 9, 2014.

[56] Interview with a USCENTCOM official, June 16, 2014.

[57] Interviews with CEWPO officials, August 21, 2014.

contains only one CCMD at present, indicating that the forces shaping the current utilization of DoD civilian deployment capabilities may *not* represent those that will generate future requirements for this capability. To develop a viable practice of civilian deployment over the long term, DoD should therefore consider the potential needs of other CCMDs beyond USCENTCOM for expeditionary civilian capabilities and, relatedly, how to broaden CCMD utilization of DoD expeditionary civilian capabilities.

Strategies for Broadening CCMD Use of DoD Expeditionary Civilian Capabilities

One way to increase the use of expeditionary civilian capabilities across the CCMDs may be to market the capability to the various commands, as detailed further in Chapter Five. Interestingly, our interviews indicate that most geographic CCMDs have a limited understanding of the benefits of expeditionary civilians and are wary of the risks they perceive to be associated with their utilization. Several CCMDs indicated that they were "ignorant" regarding the capabilities offered by expeditionary civilians or "have gotten by for this long without them."[58] Six interviewees, including officials from USCENTCOM, OSD, and the Joint Staff, noted that raising awareness of the benefits of expeditionary civilian capabilities among the CCMDs would help remedy this problem. One interviewee noted that one reason underlying USCENTCOM's extensive use of the CEW was that the capability was actively marketed to USCENTCOM officials at conferences and other events when it was first established.[59]

In addition to not fully understanding expeditionary civilian capabilities or realizing the benefits that expeditionary civilians might bring to their operations, there are indications that one reason for the CCMDs' lack of utilization of expeditionary civilians beyond the USCENTCOM AOR may be a perception of substantial risk associated with relying on civilians for deployed functions. In particular, a

---

[58] Interview with a USSOUTHCOM official, September 9, 2014; interviews with USEUCOM and USPACOM officials, September 22, 2014.

[59] Interview with a USCENTCOM official, July 29, 2014.

number of interviewees across the CCMDs, OSD, and the Joint Staff highlighted concerns regarding the *reliability* of civilians as an expeditionary workforce. There are two related components to this argument: a strongly held belief that civilians cannot be compelled to deploy and the fact that virtually all CEW personnel who have deployed to date have been volunteers (with CEW personnel with E-E and NCE designations not being used in recent years, at least not in any systematic sense).

With regard to the inability to compel a civilian to deploy, one interviewee noted, "Civilians are not as reliable as the military."[60] Yet yet another stated, "No JMD holder will write in a requirement to deploy or source a civilian because they can't force civilians to deploy."[61] Others focused on the fact that deployed CEW personnel to date have primarily volunteered for their positions; one said, "It is impossible to set up rotational plans if you have to rely on volunteers."[62]

To facilitate CCMD utilization of expeditionary civilians, the practice of civilian deployment must be structured in a manner that will allow it to address these concerns. Chapter Five presents recommendations for restructuring the practice of civilian deployment along these lines. However, we first turn to lessons learned from analogous organizations both in the United States and abroad that deploy civilians. In doing so, we reveal the contours of a viable civilian deployment structure, and we begin to see the potential benefits and drawbacks of various courses of action for DoD as it seeks to solidify its practice of civilian deployment into a feasible, sustainable long-term strategy.

---

60 Interview with a Joint Staff official, July 18, 2014.

61 Interview with a senior Joint Staff official, June 17, 2014.

62 Interview with an Air Force official, September 15, 2014.

CHAPTER FOUR

Alternative Civilian Deployment Models

As noted in Chapter One, one of the predominant research questions shaping this study entailed an exploration of the most useful lessons from alternative civilian deployment models to inform the practice of DoD civilian deployment. To derive these best practices from analogous organizations, we interviewed representatives from several DoD Fourth Estate agencies that have their own well-established civilian deployment programs, as well as officials from DoS, the U.S. Department of Homeland Security, USAID, and defense organizations in the United Kingdom, Canada, Australia, and the European Union. These interviews uncovered a wealth of knowledge regarding different possible deployment practices and the circumstances under which they are most likely to be effective. The findings presented in this chapter therefore move beyond any existing research on DoD expeditionary civilians, providing informed, robust indicators of the benefits and challenges of potential deployment models that could be used to structure a future DoD civilian deployment practice.

In analyzing the deployment processes of these various organizations, as well as the reasoning underlying their deployment mechanisms, this chapter proceeds as follows: The next section details the organizations interviewed and how they were selected for analysis. We then provide an overview of the key characteristics of the cases explored here—the relevant authorities, requirements, and mission sets pertaining to this universe of cases, as well as summary facts pertaining to the cases themselves. The bulk of the comparative analysis focuses on the development and assessment of a typology of four deployment models. The chapter concludes with common lessons learned from the four deployment models and a discussion of the relevance of these lessons to DoD's future civilian deployment practices specifically.

Selecting Cases of Civilian Deployment Analogous to the DoD Experience

For our comparative case studies, we first conducted a brief literature review that provided a general understanding of civilian deployments beyond DoD's practices, the requirements that feed deployments, and how those requirements are sourced. We then developed a list of U.S. and foreign governmental agencies that deploy civilians to at least some extent. In selecting cases for inclusion in the analysis, we sought variation in terms of the length of agencies' experiences with civilian deployment, the numbers of civilians typically deployed, and the purposes for which civilians are deployed. One of the foremost goals in selecting cases for analysis was that the universe of cases analyzed should reflect organizations similar to DoD in at least one of these respects.

In conducting this research, we found that each organization interviewed had unique missions and challenges; as a result, there were a variety of methods that organizations used to deploy their personnel. While some organizations had a narrowly focused mission set, others were responsible for a wide-ranging set of missions. To accurately reflect this variation, we ultimately made the decision to interview a set of organizations representing a diverse workforce covering a variety of missions.

Table 4.1 lists the organizational identities and types of personnel with whom we conducted interviews. Our data collection sample consisted of interviews with 17 governmental agencies, both inside and outside the United States. These included DoD Fourth Estate agencies that have their own unique civilian deployment processes, such as DLA, DIA, and DCMA; the Bureau of Conflict Stability and Diplomatic Security Office in DoS; the Office of International Affairs, CBP, DEA, and FEMA in the U.S. Department of Homeland Security; and USAID's CS3 and OTI. In terms of foreign analogous organizations,

Table 4.1
Organizations Interviewed for This Study

Agency Type	Office	Number of Interviewees
U.S. government agencies		
DoS	Bureau of Conflict and Stabilization Operations	1
	Bureau of South and Central Asian Affairs	1
	Diplomatic Security Office	1
	APSPO	1
USAID	CS3	1
	OTI	1
U.S. Department of Homeland Security	Office of International Affairs	1
	CBP	2
	FEMA	1
U.S. Department of Justice	DEA	2
DoD Fourth Estate	DIA	5
	DLA	1
	DCMA	3
Foreign government agencies		
UK MOD	S2O	2
Canadian Department of National Defence	J1 (personnel) and human resources	2
Australian Department of Defence	DSTO	1
EEAS	CPCC	6
	CPCC Civilian Response Teams	1
Total		33

NOTES: The 33 interviewees listed here are those from analogous U.S. and foreign organizations. They do not include interviewees from OSD, the Joint Staff, the CCMDs, or the U.S. military. Altogether, we interviewed 83 individuals from 45 offices for this research, including those listed here.

we conducted interviews with officials from the UK MOD's S2O team, the Canadian Department of National Defence, the Australian DSTO (now the Defence Science and Technology Group), and the European Union's EEAS CPCC. We interviewed a total of 33 personnel from across these organizations.

Throughout the interview process, we accumulated a wealth of knowledge regarding specific civilian deployment experiences, including the requirements that generate the need for deployable civilians, the types of missions they support, and the methods that organizations use to identify, select, track, and deploy civilians. During our interviews, it became apparent that organizations use many different methods to deploy personnel. These findings and good practices are outlined throughout this chapter.

Overview of Case Characteristics

Defining Key Terms and Concepts

Authorities

Civilians routinely deploy to support missions through a variety of authorities. Deployed civilians from the U.S. government agencies examined typically operate either under chief of mission (COM) authorities or through DoD under authorities derived from Title 10 of the U.S.C. Under most circumstances, deployed personnel are ultimately the responsibility of either the U.S. ambassador or a military commander.

Most civilian U.S. agencies deploy their personnel to a contingency operation under COM authority. The DoS *Foreign Affairs Manual, Volume 2, Handbook 2*, clearly describes the COM's authority and the processes to exercise it over U.S. government staffing and personnel for missions abroad. "COMs are the principal officers in charge of U.S. Diplomatic Missions and certain U.S. offices abroad that the Secretary of State designates as diplomatic in nature. The U.S. Ambassador to a foreign country, or the *chargé d'affairs*, is the COM in that country."[1] A number of documents provide guidance and the legal basis for these authorities, including the President's letter of instruction to COMs, the DoS Basic Authorities Act, the 1980 Foreign Service Act, the 1986 Diplomatic Security Act, and National Security Decision Directive 38.[2] The COM has authority over every executive branch agency in a host country but not personnel under the command of a U.S. military commander—typically the combatant commander or geographic combatant commander—or those on the staff of an international organization.

Other relevant authorities are derived from Title 10 and are inherently military in nature. Combatant commanders consider strategic guidance documents, including the *Unified Command Plan, National Security Strategy, National Defense Strategy, National Military Strategy, the Quadrennial Defense Review Report, Guidance for the Development of the Force, Guidance for Employment of the Force,* and *Joint Strategic Capabilities Plan,* in selecting forces for various missions.[3] The commander then requests forces through the Joint Chiefs of Staff, who, in turn, make their recommendations to the Secretary of Defense. The Secretary of Defense then assigns, apportions, or allocates military and civilian personnel to CCMDs for mission execution. Typically, civilians will be employed under one of these authorities in a contingency operation.

---

1 U.S. Department of State, *Foreign Affairs Manual*, 2 F.A.H.-2 H-112, "Foreign Assistance Standardized Program Structure and Definitions," April 8, 2010.

2 For a full review of these documents, see U.S. Department of State, *Foreign Affairs Manual*, 2 FAH-2 H-100, "Post Management Organization and Chief of Mission Authority and Overseas Staffing," July 18, 2014.

3 See, for example, Executive Office of the President, *National Security Strategy*, Washington, D.C.: White House, May 2010; U.S. Department of Defense, *Sustaining U.S. Global Leadership: Priorities for 21st Century Defense*, Washington, D.C., January 5, 2012; U.S. Joint Chiefs of Staff, *The National Military Strategy of the United States of America: Redefining America's Military Leadership*, Washington, D.C., February 2011a; DoD, 2006; U.S. Department of Defense, *Quadrennial Defense Review Report*, Washington, D.C., February 2010; and U.S. Department of Defense, *Quadrennial Defense Review 2014*, Washington, D.C., March 4, 2014.

Due to the increased terrorism threat and the need to ensure the security of U.S. personnel and facilities internationally, the Secretary of State and Secretary of Defense signed a memorandum of understanding with the effective date of December 16, 1997 to establish these security responsibilities.[4] Pursuant to this memorandum, responsibility for security has been clearly delineated between DoS and DoD through a series of individual country agreements that assign responsibility for security of DoD personnel in a given country to either the COM or the combatant commander, depending on the mission. The memorandum has served to resolve confusion about security responsibilities.[5]

The COM authority and other Title 10 authorities cover U.S personnel only. As noted, COM authority typically covers civilian agencies, and Title 10 authorities cover personnel assigned to CCMDs. Other countries reviewed for this study have different rules and regulations that govern the employment of civilians. However, while each country has its own legal rules and regulations that must be satisfied to deploy personnel, most of the organizations that we examined do have a similar framework for civilian deployment. Similarly, for deployments in the context of European Union Common Security and Defence Policy (CSDP) civilian missions, the civilians deploy under an EU mandate approved by the Council of the European Union.

Requirements

Requirements drive operational missions. For the purposes of this study, we viewed requirements as the set of activities necessary to develop, consolidate, coordinate, validate, approve, and prioritize the deployment of civilian personnel to contingency operations. Many of the agencies included in this study have adapted portions of their organizations, business processes, and deployment models to address requirement requests. The origin of the majority of the requirement requests determines the type of deployment model an agency uses. There are

---

[4] "Memorandum of Agreement Between the Departments of State and Defense on the Protection and Evacuation of U.S. Citizens and Nationals and Designated Other Persons from Threatened Areas Overseas," July 1998.

[5] DoS, 2010.

two typical methods through which a request can come to an agency: top-down or bottom-up.

Top-down requests originate from division, bureau, or secretariat/headquarters levels, the National Security Council, or congressional or presidential direction. Top-down requests are primarily directive in nature and compel the organization to react to the request. Bottom-up requests, on the other hand, come to an agency from a variety of sources outside the organizational chain of command and could require either immediate attention or a delayed response. In this case, the requesting agent often makes the request through a U.S. embassy on behalf of a partner nation or the request comes directly from embassy staff via the COM. Requests can also come to an agency through other federal agencies or through such international organizations as NATO, the European Union, and the United Nations.

The organizations in our study must balance efficiency and speed with personnel identification. In general, the speed with which a requirement needs to be filled will determine whether the individual selected for a deployment should come from within the organization or hired from outside the organization, as well as whether a preselected pool of candidates is needed prior to requirement identification.

Mission Types

There are a variety of missions that civilian agencies routinely deploy personnel to support, ranging from relatively benign workshops and technical assistance programs to efforts aimed at countering extremist operations in high-threat environments. The agencies interviewed for this project covered a host of non–steady-state operations that require civilian expertise, including humanitarian assistance/disaster relief, stabilization and reconstruction, counterdrug operations, counterpiracy operations, capacity building, institution building, election monitoring, intelligence, countergang operations, technical assistance, liaison and coordination duties, and security force training.

While military personnel can and do conduct many of these missions, in many cases, specific civilian expertise is desired. Military personnel are often considered "generalists" outside of their core warfighting functions. On the other hand, civilians are expected to be experts and may have more in-depth knowledge on a particular topic. The key is discerning when to leverage civilian expertise or when a generalist will suffice. Many of the skills required for the mission types listed previously center on specific expertise found predominantly in civilian agencies.

Case Overview

Next, we present a brief summary of each of the organizations that we examined for this study. While we mention deployment numbers obtained through the interview process, it is important to note that because our focus was on the *process* of deployment, we did not comprehensively analyze and cross-check the numbers of personnel deployed by each organization. We did find, however, that the number of civilians deployed by organizations tended to vary on an annual basis depending on missions and requirements. Furthermore, not all organizations were in a position to give an exact number of annual deployees. We include the estimated numbers in this report to give a rough indication of the size and scope of each organization's civilian deployments in an effort to demonstrate the analogy between these organizations and DoD.

Table 4.2 captures many of the findings from our interviews. The "Deployment Type" column indicates whether requirements are part of steady-state operations or are typically emergent requests. The "Deployment Office" column indicates the structure of the office that deploys civilians: Centralized offices maintain more oversight of deployed personnel, and decentralized offices relinquish control. The "Requirement Source" column indicates where most requirements are generated. Finally, "Sourcing" indicates the source of personnel used to fulfill deployable civilian requirements.

Bureau of Conflict and Stabilization Operations, U.S. Department of State

The Bureau of Conflict and Stabilization Operations "advances U.S. national security by breaking cycles of violent conflict and mitigating

Table 4.2
Characteristics of Organizations Interviewed

Organization Type	Annual Number	Deployment Type	Deployment Office	Requirement Source	Sourcing	Volunteers
U.S. government agencies						
DoS						
Bureau of Conflict and Stabilization Operations	30	Both short-notice and planned	Decentralized office	COM, functional bureaus, CCMD	Through embassy or regional or functional bureau; identify need through the Crisis Response Network	Yes
Bureau of South and Central Asian Affairs	Declined to comment	Both short-notice and planned	Centralized office	COM, functional bureaus, CCMD	Through embassy or regional or functional bureau	—
Bureau of Diplomatic Security	Declined to comment	Both short-notice and planned	Centralized office	Embassy	Internal to the bureau	Yes
APSPO	50–100	Planned	Centralized office	Embassy	Hire externally for most positions	Yes
USAID						
CS3	50	Both short-notice and planned	Decentralized office	Embassy	Bullpen	Part of job
OTI	190	Both short-notice and planned	Decentralized office	Embassy	Bullpen	Part of job

Table 4.2—Continued

Organization Type	Annual Number	Deployment Type	Deployment Office	Requirement Source	Sourcing	Volunteers
U.S. Department of Homeland Security						
Office of International Affairs	?	Both short-notice and planned	Decentralized office	Embassy	Internally for most positions	Yes
CBP	750	Both short-notice and planned	Centralized office	Embassy	Volunteers and internal staffing	Yes
FEMA	4,000	Short-notice	Centralized office	National Response Coordination Center	Via executive office and a declared emergency	Yes
U.S. Department of Justice						
DEA	800 positions overseas	Both short-notice and planned	Centralized office	Embassy, long-standing offices	Lengthy internal process	Yes
DoD Fourth Estate						
DIA	100–150	Both short-notice and planned	Centralized office	CCMD, GFM	Through mission managers within each directorate, internally	Part of job
DLA	200–300	Both short-notice and planned	Centralized office	CCMD, GFM	Volunteers and internal staffing	Yes
DCMA	50–100	Planned	Centralized office	CCMD, GFM, joint task force, forward-stationed contract management office	Moving to be a source provider, not an executor; the services will execute their contracts	—

Table 4.2—Continued

Organization Type	Annual Number	Deployment Type	Deployment Office	Requirement Source	Sourcing	Volunteers
Foreign government agencies						
UK MOD						
S2O	150	Both short-notice and planned	Centralized office	Theater or permanent joint headquarters or elsewhere	From MOD civil servants; sometimes the wider UK civil service	Yes
Canada						
Department of National Defence (J1)	100	Both short-notice and planned	Centralized office	Theater	Within the pool of existing public servants	Yes
Australia						
DSTO	10	Both short-notice and planned	Centralized office	Operational commanders	Within DSTO or Australian Department of Defence	Yes
EEAS						
CPCC	3,200 currently deployed	Both short-notice and planned	Centralized office	Committee for Civilian Aspects of Crisis Management	Within the member state; alternatively, third states if required	Yes
CPCC Civilian Response Teams	—	Short-notice	Centralized office	Committee for Civilian Aspects of Crisis Management	Within the member state; alternatively, third states if required	Yes

crises in priority countries."[6] It falls under the purview of the Office of the Under Secretary of State for Civilian Security, Democracy, and Human Rights and was created by Secretary of State Hillary Clinton in 2012 to improve the U.S. response to conflicts and crises in other countries. Its missions often require civilian involvement and expertise, and it has historically deployed around 30 personnel a year.[7] The bureau has also created a pool of civilian experts with specific characteristics who can be alerted and deployed on short notice, typically within two weeks.[8]

Bureau of Diplomatic Security, U.S. Department of State

Bureau of Diplomatic Security is responsible for security and law enforcement within DoS. Inside the United States, it is responsible for the protection of the Secretary of State and visiting high-ranking dignitaries and other visiting officials. Overseas, it provides personnel and embassy security in more than 160 foreign countries across 275 U.S. diplomatic missions. It can deploy personnel as individuals or in a variety of teams, including security support teams, tactical support teams, and mobile training teams. It leads international investigations into passport and visa fraud, conducts personnel security investigations, and assists in threat analysis, cyber security, and counterterrorism missions.[9]

Crisis Surge Support Staff and Office of Transition Initiatives, U.S. Agency for International Development

Both CS3 and OTI deploy personnel with the necessary skill sets to further U.S. foreign interests with the goal of improving lives and livelihoods in the developing world. One of the distinguishing characteristics of these organizations is the ability to provide a surge capability to

---

6 U.S. Department of State, "Bureau of Conflict and Stabilization Operations," web page, undated.

7 Interview with a Bureau of Conflict and Stabilization Operations official, September 18, 2014.

8 Interview with a Bureau of Conflict and Stabilization Operations official, September 18, 2014.

9 U.S. Department of State, "Bureau of Diplomatic Security," web page, undated.

U.S. missions through a flexible and quick-reaction deployment mechanism that selects, screens, trains, and holds individuals in a wait status until a requirement emerges. Individuals in this wait status are said to be "on the bench" or "in the bullpen." Each office deploys between 50 and 190 personnel annually.[10] While CS3 tends to focus directly on short-term U.S. embassy support, OTI works primarily through implementation partners to quickly provide goods and services in crisis situations.[11]

U.S. Customs and Border Protection, U.S. Department of Homeland Security

CBP is charged with securing more than 7,000 miles of U.S. land borders and 328 ports of entry. It is responsible for protecting American citizens from terrorist threats and preventing the illegal entry of persons and goods.[12] CBP also facilitates the lawful travel and trade of goods and services across U.S borders. It has more than 42,000 officers and border-control agents who are deployed throughout the United States. Outside the United States, more than 750 agency personnel operate under COM authority in a variety of roles, including as attachés, advisers, representatives, and security personnel in support of specific missions and programs. CBP requirements are generated through a variety of multiyear initiatives, as well as quick staffing solutions to fulfill short-term, ad hoc needs.[13] To fill these latter requirements, CBP has developed a database of prescreened personnel, centered on a core group of 22 staff who can conduct short-notice training events.[14]

Federal Emergency Management Agency, U.S. Department of Homeland Security

FEMA's primary role is to coordinate the response of federal, state, and local authorities in the event of a natural or man-made disaster.

---

[10] Interviews with OTI and CS3 officials, September 15, 2014.

[11] Interviews with OTI and CS3 officials, September 15, 2014.

[12] U.S. Customs and Border Protection, "About CBP," web page, undated(a).

[13] U.S. Customs and Border Protection, undated(a).

[14] Interviews with CBP officials, September 18, 2014.

The organization has around 23 directorates, ten regional operations centers, and an incident management and support staff of more than 17,000 personnel.[15] It has a tiered approach to readiness that allows some disaster response experts to deploy quickly while simultaneously notifying other FEMA employees of the disaster and that they might be required to deploy. For example, the incident management staff are full-time, fully trained FEMA employees who respond immediately in the event of a disaster. Depending on the severity of an event, ancillary support personnel can be called to help augment the incident management staff. Ancillary support can come from local, state, or other directorates within FEMA. The FEMA Corps, a cadre of 18- to 24-year-olds dedicated to disaster response, is one such organization that can be used in a disaster. It consists of a small number of highly skilled disaster assistance operators and is kept in a high state of readiness to deploy on short notice.[16] The U.S. Department of Homeland Security Surge Capacity is another standby force. It consists of 4,000 federal employees who can be called in the event of an emergency to provide additional capability to FEMA.[17]

Drug Enforcement Administration, U.S. Department of Justice

DEA's mission is "to enforce the controlled substances laws and regulations of the United States and bring to the criminal and civil justice system of the United States, or any other competent jurisdiction, those organizations and . . . members of organizations, involved in the growing, manufacture, or distribution of controlled substances."[18] DEA has roughly 800 positions overseas, representing approximately 10 percent of its workforce, and has been deploying agents and support staff to overseas missions under COM authority for more than 25 years.[19] Deployed personnel support a variety of missions and activities, including the management of a national drug intelligence program, investiga-

---

[15] Interview with a FEMA official, September 19, 2014.

[16] Interview with a FEMA official, September 19, 2014.

[17] Interview with a FEMA official, September 19, 2014.

[18] Drug Enforcement Administration, homepage, undated.

[19] Interviews with DEA officials, September 24, 2014.

tion and preparation of cases for prosecution, liaison and coordination duties, training activities, and investigative and strategic intelligence gathering.

Defense Intelligence Agency, U.S. Department of Defense

DIA has deployed a range of operational and support personnel since the Vietnam War, but it was not until 2002 that it began emphasizing civilian deployment. Personnel routinely deployed since 2002 include analysis, collection, IT support, logistics, administrative, finance, and contracting officers.[20] Civilian deployment requirements vary by year, with current requirements hovering around 100–150 billets.[21] DIA's Expeditionary Readiness Center provides training, administrative, and medical support to deploying personnel. The center also provides many of the same services to other Intelligence Community organizations through memoranda of agreement or understanding, including the National Geospatial-Intelligence Agency, the National Reconnaissance Office, the Office of the Director of National Intelligence, and the National Security Agency. Like DLA and DCMA, DIA is part of the DoD Fourth Estate.

Defense Logistics Agency, U.S. Department of Defense

"As America's combat logistics support agency, the Defense Logistics Agency provides the Army, Navy, Air Force, Marine Corps, other federal agencies, and combined and allied forces with the full spectrum of logistics, acquisition and technical services."[22] DLA employs approximately 27,000 personnel, of whom 1,000 are military and the rest civilian. It has personnel stationed overseas at distribution centers in support of routine missions, but the agency has also deployed up to 300 civilian staff in support of contingency operation requirements.

---

[20] Defense Intelligence Agency, "About DIA," web page, undated.

[21] Interviews with DIA officials, October 10, 2014.

[22] Defense Logistics Agency, "At a Glance," web page, undated.

Defense Contract Management Agency, U.S. Department of Defense

DCMA consists of more than 11,900 civilians and military personnel who manage the execution of contracts on behalf of DoD that cover more than 20,000 contractors.[23] Although the agency was established only in 2000, it has undergone significant change—from a primarily domestic contract oversight role to that of an expeditionary force provider. In that capacity, DMCA at one time deployed up to 450 civilian contracting and support personnel, but with the subsequent drawdown of forces, current requirements range from 50 to 100 deployed personnel.[24]

UK Ministry of Defence Support-to-Operations Team

MOD's S2O office was established in 2006 to enable the generation, deployment, and subsequent redeployment of MOD civilians in support of overseas operations. Its policy and communication team is responsible for deployment policy, rules, and guidance, as well as promoting the program and managing information disseminated to the S2O community.[25] The safety and security team is responsible for managing the risks associated with deploying to operational theaters and for the policies concerning safety, security, and visits. Finally, the administrative support team handles the administrative elements of deployments, including booking flights, processing operational allowances, and scheduling individuals for training. The roles that this team supports include policy advisers, civil secretaries, media advisers, and operational analysts.[26] Each role has a designated senior-level official who is responsible for maintaining pools of volunteers for deployment. For some roles, this also includes high-readiness pools; however, these are currently in an early stage of development.

---

[23] Defense Contract Management Agency, "About the Agency," web page, undated.

[24] Interviews with DCMA officials, September 11, 2014.

[25] Interview with an S2O official, August 2014.

[26] Interview with an S2O official, August 2014.

Canadian Department of National Defence

Since its involvement in Afghanistan in 2001, the Canadian Department of National Defence has deployed civilian specialists to operational theaters. Personnel deployed include medical specialists, morale and welfare staff, policy advisers, and intelligence analysts.[27] While many individuals deploy under the public service umbrella, some are sourced through the Canadian Department of Foreign Affairs. The civilian requirement varies across missions but has entailed an average of 80 deployees working on the ground in Afghanistan.[28]

Australian Defence Science and Technology Organisation

DSTO, which is now known as the Defence Science and Technology Group, consists of approximately 2,300 civilian staff employed as scientists, engineers, IT specialists, and technicians.[29] It is part of the Australian Department of Defence and supports scientific analysis and research and development. As part of this mission, the organization deploys scientists in support of military operations to provide immediate, on-the-ground advice and assistance. Personnel deployed in these roles include operational analysts, anthropologists, and cultural advisers.[30] Scientists are paired with military personnel and deploy as a team. This pairing is established during predeployment training and continues throughout the deployment. DSTO's requirement for particular civilian skill sets has varied over time, from geospatial specialists to analysts skilled in developing metrics to understand strategic impact.[31]

---

27 Interviews with Canadian Department of National Defence officials, September 2014.

28 Interviews with Canadian Department of National Defence officials, September 2014.

29 Defence Science and Technology Group, "About DST Group," web page, undated. DSTO changed its name since the time that this research was conducted. We refer to the organization by its name at the time of our study because we did not investigate whether there were attendant structural changes to the organization.

30 Interview with a DSTO official, September 25, 2014.

31 Interview with a DSTO official, September 25, 2014.

Civilian Planning and Conduct Capability, European External Action Service

CPCC, part of the EEAS supports the sourcing of staff to CSDP missions. The first such mission was launched in 2003. Since then, the European Union has launched 24 civilian missions and military operations.[32] In 2013, CPCC supported ten civilian missions, including training missions, border and judicial system support, support for security-sector reform, support to authorities in combating terrorism and organized crime, and more general advice or assistance with defense reform. Deployed personnel come from EU member states and third-party states (those outside of the European Union).[33] The missions range in duration, depending on the mission mandate. Missions involve the deployment of roughly 3,200 military personnel, and CPCC had around 3,700 civilians deployed at the time of this research. The Civilian Headline Goal 2010 aimed to improve the European Union's civilian capability to respond effectively to crisis management tasks in the context of CSDP.[34] One focus was on improving the capabilities and capacity of civilians, for instance through improved predeployment training. In 2011, Europe's New Training Initiative for Civilian Crisis Management (known as ENTRi) was launched to prepare and train crisis management personnel in a rigorous and standardized manner.[35]

Typology of Deployment Models

Overall, we identified four models that these organizations have applied to deploy civilians. The models differ along two main dimen-

---

[32] European Parliament, Directorate General for External Policies of the Union, Policy Department, *CSDP Missions and Operations: Lessons Learned Processes*, Luxembourg: European Union Publications Office, April 2012.

[33] Interviews with CPCC officials, October 10, 2014.

[34] "Civilian Headline Goal 2010," approved by the ministerial Civilian Capabilities Improvement Conference and noted by the General Affairs and External Relations Council, Document 14823/07, November 19, 2007.

[35] ENTRi, homepage, undated.

sions: the extent to which they sourced individuals to deploy from within the organization's existing civilian ranks (internal sourcing), as opposed to searching for candidates external to the organization (external sourcing), and the extent to which the organizations had a pool of preidentified individuals before the issuance of requirements (proactive sourcing), as opposed to identifying candidates for positions after requirements had been issued (reactive sourcing). Table 4.3 categorizes the 17 analogous organizations into the four deployment sourcing models.

It is important to note that some of these organizations can be classified into more than one category, depending on the office in the organization in which it is housed. We therefore categorized each orga-

Table 4.3
Deployment Models, by Agency

	Reactive Sourcing	Proactive Sourcing
Internal Sourcing	CBP DEA DIA Canadian Department of National Defence DoS Bureau of South and Central Asian Affairs U.S. Department of Homeland Security Office of International Affairs	DCMA DLA FEMA DoS Bureau of Diplomatic Security UK MOD S2O Australian DSTO (now the Defence Science and Technology Group)
External Sourcing	APSPO EEAS CPCC	USAID OTI USAID CS3 DoS Bureau of Conflict and Stabilization Operations EEAS CPCC, Civilian Response Teams

nization based on the *predominant* sourcing model it used to fill the *majority* of civilian requirements.[36]

Reactive Internal Sourcing

We termed the first model *reactive internal sourcing* (see Figure 4.1). In this model, a requirement is identified through either a top-down or a bottom-up process.[37] The requirement triggers a recruitment process internal to the organization—for instance, within the civil service of a defense department. Individuals are selected to fill the requirement, after which they undertake any required predeployment training or preparation (e.g., medical screening and vaccinations, visas, clearances, cultural training, hostile environment training) as needed for the specific deployment. Individuals deploy to their posting, and after deployment they return to the post they had occupied prior to deployment.[38] Only a few organizations specifically mentioned decompression as part of the process.[39]

Benefits and Constraints

The reactive internal sourcing model is beneficial in that it entails (and enables) longer-term organizational ownership of the skills required by deployed civilians. Because it focuses solely on candidates internal

---

[36] Quick-turn, emergent requirements may necessitate deploying personnel in a manner different from the process traditionally used by an organization. From our interviews, we determined which deployment model was used most often to deploy each organization's civilian personnel. If there was deviation from that standard, it usually occurred on a case-by-case basis.

[37] The requirement in some instances had to be validated through engagement between theater and home office leadership to ensure that it was valid and that a civilian was best placed to fill it before individuals were sourced (interviews with S2O officials, August 2014).

[38] In some organizations, individuals were given a preview of life during operations to ensure that people's decisions to volunteer were based on realistic information about the position. In the literature on organizational selection, this practice is often referred to as a "realistic preview," happening prior to the application process (interviews with S2O officials, August 2014; interview with a DSTO official, September 25, 2014; interviews with DIA, DEA, and CBP officials, September and October 2014).

[39] Interviews with S2O officials, August 2014; interview with a DSTO official, September 25, 2014; interviews with Canadian Department of National Defence officials, September 2014; interviews with DIA officials, October 10, 2014.

Figure 4.1
Reactive Internal Sourcing Model

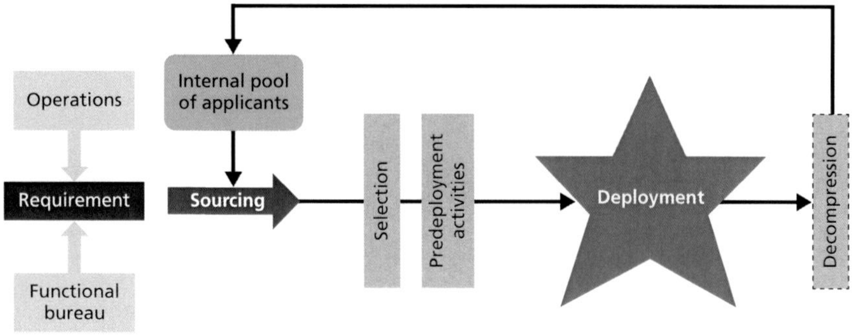

RAND *RR975-4.1*

to the organization, it ensures that the organization maintains these civilian capabilities within its overall workforce following any particular deployment. Personnel who are deploying complete training and preparation just prior to deployment, such that costs are not incurred well in advance of deployment and the training can be targeted for the specific deployment. In many cases, the deployment training and additional skills developed by civilians while deployed have added benefits that can be applied back to the home organization upon return. The individuals involved with the candidate selection process—handled by management or a selection board that is familiar with the requirement—understand the organization, mission, and capabilities and are selected according to the specific requirement.[40]

However, there are also a number of potential drawbacks associated with this model. The length of the process means that it is not well suited for short-notice, urgent deployments (unless there is a speedy internal process for advertising and recruiting staff on short notice).[41] Applicants are also already employed in other roles within their organization, which means that when they deploy, their posts are often left open without backfill. Additionally, for certain skill sets, the candidate

---

[40] Interviews with DIA officials, October 10, 2014.

[41] Interviews with CPCC officials, October 10, 2014.

pool may be too shallow under this model, causing a capability shortage if an individual with the necessary qualifications does not apply to fill the requirement.[42] The more complex the expertise needed, the more difficult it is to find suitable candidates.[43]

Other constraints include issues with reintegrating deployed personnel back into the home office. Sometimes, individuals do not want to return to their previous role or position due to their newly acquired experience.[44] Other times, employees are penalized for deploying, or there is home office animosity toward the deployed person because his or her position was gapped without backfill. Personnel have returned to find that their former jobs have been filled.

Finally, some interviewees identified issues surrounding traditional human resources functions. Several challenges arose concerning the identification and selection of potential deployed personnel, performance evaluations, and the overall flow of information from the human resources office to potential deployees. Other interviewees raised the issue of unfair promotion practices.[45]

Proactive Internal Sourcing

The second model identified through our analysis is the *proactive internal sourcing model* (see Figure 4.2). Similar to reactive internal sourcing, under this model deployed civilians are sourced from within an organization's existing civilian employee pool. Under this model, there are two methods that organizations use to source personnel for the readiness pool. In the first method, organizations stipulate that the applicant must be able to deploy as a condition of employment. In this

---

[42] For instance, since 2006, there have been 43 calls for contribution to CSDP civilian missions, with a goal of filling 150 posts. Of these posts, 109 were filled through this process, though only 23 were filled with candidates from the expert pool (interviews with CPCC officials, October 10, 2014). For selected organizations that have maintained a high deployment tempo over multiple years, the candidate pool is sufficiently deep, but candidate availability has been diminished by consecutive deployments.

[43] Interviews with CPCC officials, October 10, 2014.

[44] Interviews with S2O officials, August 2014.

[45] Interviews with DIA, DEA, and CBP officials, September 2014 and October 2014.

Figure 4.2
Proactive Internal Sourcing Model

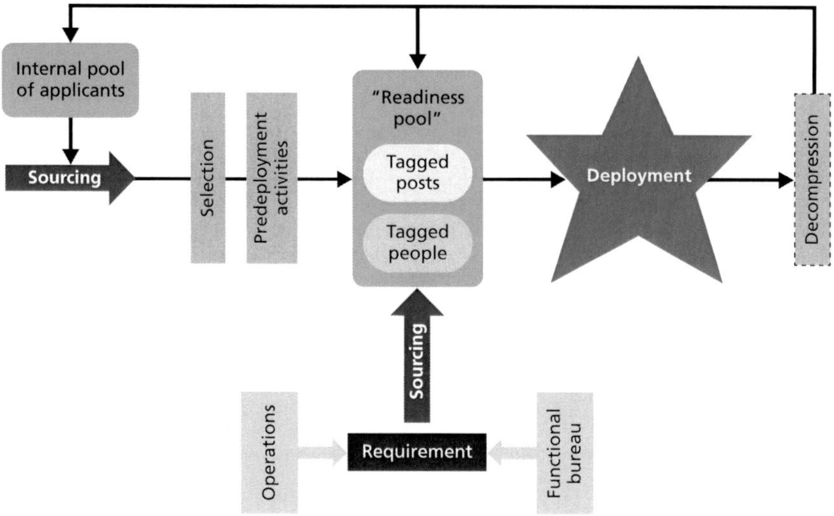

RAND *RR975-4.2*

case, personnel are hired into a "tagged post" that identifies personnel hired to fulfill a specific position. Personnel can be hired to fill a deployment billet, or to fill a home station billet with the condition that they are required to deploy as a routine part of their job. In the second method, an organization can hire personnel for a home station position and then fill its readiness pool with volunteers. Once a person volunteers, he or she becomes a "tagged person," is placed into the readiness pool, and can be called on to deploy. Whereas under the first method an offer of employment is contingent upon agreeing to fill a tagged post, the second method relies on volunteers. The manner in which the readiness pool is filled varies by organization, with some organizations hiring personnel only into tagged posts, others using only volunteers via the tagged person method, and still others using a combination of the two methods.

However the readiness pool is sourced in this model, organizations do not wait for a specific requirement to be identified. Rather, an organization will preidentify a set of the most probable requirements and use those as the basis for establishing a readiness pool.

The organization will then use its own civilian employees, identified by either tagged post or tagged person, to source the set of emergent requirements. Before being placed into the readiness pool, individuals undergo required predeployment training and screening to ensure that they are able to deploy. Upon completion, they are placed into the readiness pool.

Benefits and Constraints

The primary benefit of the proactive internal sourcing model is the ability to deploy on relatively short notice, because individuals are already preselected and prepared for a set of likely missions. Similar to the previous model, the organization retains ownership of the skills needed with regard to deployment. Furthermore, there is an opportunity in both internally sourced models for organizations to learn from the experience and to retain the expertise of deployed civilian personnel upon their return.

As with the previous model, drawbacks associated with proactive internal sourcing include a lack of backfill for (and the requirement to hold open) home office postings, as well as related difficulties as the civilian attempts to reintegrate into the home office after deployment. Individuals in the readiness pool are not guaranteed to deploy. For example, the requirement might never emerge, it might take too long for the requirement to emerge, or there could be an issue with retaining personnel in the pool.[46] Multiple organizations that utilize this model encourage the entities responsible for deploying civilian personnel to actively monitor their pools to ensure that they are appropriately sized to meet requirements and that personnel are not in a pool for so long that they lose interest in deploying.[47] The cost of predeployment training is also incurred regardless of deployment, and civilians in the readiness pool may need refresher training or new training, depending on how much time has elapsed since their recruitment into the pool.

A final drawback of the proactive internal sourcing approach is that the forecasted requirement against which the individuals were

---

[46] Interview with a DSTO official, September 24, 2014

[47] Interviews with S2O officials, August 26, 2014.

originally recruited may evolve and differ from actual future requirements.[48] Therefore, a process must be established to routinely validate the set of requirements and the appropriateness of the skill sets represented in the readiness pool. For example, some agencies have a quarterly validation panel that looks at current and future requirements; the readiness pool is subsequently adjusted according to these new requirements.[49]

Reactive External Sourcing

The third model identified in this comparative analysis is the *reactive external sourcing model* (see Figure 4.3). In this model, organizations draw individuals from external sources for deployable civilian positions. That is, a specific requirement is identified, the organization advertises the requirement externally, and an outside expert is hired to fill the requirement. The organization usually covers the costs associated with any necessary training, medical screening, visa, and security clearances for the individual in question. Following deployment, the employees hired under this model are no longer affiliated with the organization,

Figure 4.3
Reactive External Sourcing Model

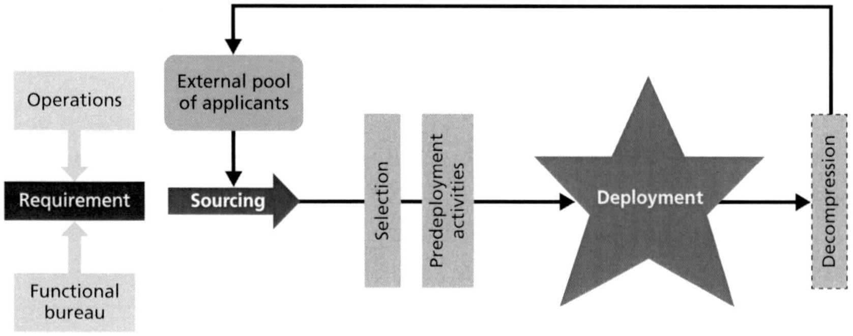

RAND *RR975-4.3*

---

[48] Interview with a Canadian Department of National Defence official, September 2014; interviews with CPCC officials, October 10, 2014.

[49] Interviews with officials from analogous civilian deployment organizations, July–November 2014; interviews with U.S. government officials, 2014.

often returning to their former posts with other organizations (including universities). APSPO in DoS is one organization that utilizes this approach.[50] We also categorized CPCC within this category, though the distinction between internal and external sourcing in CPCC is less clear-cut because a call for contributions is sent out to member states, which then look internally to their government departments for candidates. Once a candidate is selected, he or she will fill that requirement and deploy for the mission. If a call for contributions has been sent out twice without enough volunteers, the call is expanded to third states.

Benefits and Constraints

The benefits of the reactive external sourcing model are similar to those of the reactive internal sourcing model, particularly with regard to targeting predeployment training only to those who will be deployed. The selection procedure is focused on finding the most highly qualified individuals matching the requirement. Because individuals are sourced externally, there is no issue with backfilling home office assignments within the organization, and the costs are not incurred unless individuals are actually hired to fulfill a requirement and deploy.

Challenges associated with this model include an ever-present question as to whether the skills needed for any particular requirement will be readily available in the external environment. In our interviews, this was typically not a concern.[51] However, it is possible to have a scenario in which specific requirements are hard to fill because the capability is not readily available outside the organization. Furthermore, if skills attractive to the home office are developed during deployments, they are not retained after deployment under this model, as deployees are not retained in the organization following their deployment. Often, external recruitment is a lengthy process and does not lend itself to urgent short-notice deployments because of the U.S. Office of Person-

---

[50] Interview with an APSPO official, September 10, 2014.

[51] Several interviewees mentioned the difficulty in finding personnel with the necessary skill sets for highly technical work, such as electricians, rule-of-law specialists, DNA analysts, air traffic controllers, and English-language specialists (interview with a U.S. government official, September 2014; interviews with CPCC officials, October 10, 2014).

nel Management's competitive hiring authorities.[52] Finally, additional training or security requirements associated with deploying external candidates may need to be considered.

Proactive External Sourcing

The fourth model identified through our analysis is the *proactive external sourcing model* (see Figure 4.4). Much like the reactive external sourcing model, personnel from outside of the organization are identified to fill requirements. The organization uses various planning models to forecast a set of future requirements and then hires personnel to source those requirements. In anticipation of a requirement for civilian deployment, organizations such as OTI and CS3 set up a *bullpen*—a readiness pool of external selectees used to fill requirements

Figure 4.4
Proactive External Sourcing Model

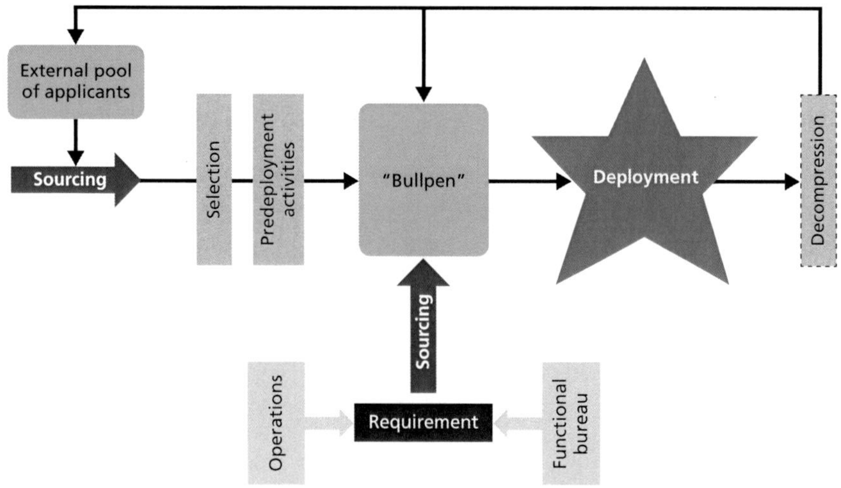

RAND *RR975-4.4*

---

[52] Several interviewees mentioned that the fair hiring practices mandated by the U.S. Office of Personnel Management require a thorough screening of all applicants, which lengthens the hiring process. See Office of Personnel Management, "Hiring Authorities: Competitive Hiring," web page, undated.

when needed.[53] A practice commonly seen in organizations utilizing this model involves selectively hiring experts prior to the issuance of actual requirements, conducting predeployment training and medical screening, obtaining passports and security clearances, and then placing candidates in the bullpen. While in the bullpen waiting to be called for a deployment, individuals are not paid nor provided benefits. When a requirement is issued that matches the qualifications of a particular individual in the bullpen, he or she is notified for a deployment and then his or her pay and benefits are activated.

Benefits and Constraints

As seen in the proactive internal sourcing model, the period between identified requirement and actual deployment is likely to be shorter in the proactive external sourcing model because individuals are already preidentified, have been selected, and have undertaken required predeployment readiness preparation. Salary-related costs are incurred only after personnel are deployed under this model and, depending on the organization, would be paid by either the home office or the field office. For example, the home office pays most salary costs for deployed personnel in OTI. Conversely, at CS3, the embassy mission that deployed personnel are supporting predominantly covers salary costs.[54]

Most organizations do not need to backfill posts due to the nature and function of the bullpen. Requirements are forecasted such that the necessary qualifications are understood in general terms, and experts with the necessary knowledge and skill sets to meet these qualifications are selected for the bullpen. Many organizations that use this model have stringent authority over the hiring and firing of personnel, and there are few bureaucratic processes associated with relieving personnel who are not a good fit.[55]

Yet, this model also has its share of drawbacks. Although salaries are paid only upon actual deployment, the organization may incur

---

53 Interviews with OTI and CS3 officials, September 2014.

54 Interviews with OTI and CS3 officials, September 15, 2014.

55 Interview with an OTI official, September 15, 2014.

the costs associated with predeployment training up front regardless of whether an individual actually deploys. If the requirement changes from what was originally anticipated, there may be issues with not having the required capability or skill set in the bullpen, and the training and readiness costs are sunk expenses that cannot be retrieved for personnel who no longer meet the qualifications of evolving forecasted requirements. Finally, as seen in the previous model, the skills developed by the individual during his or her deployment are not retained easily within the organization.

Lessons from the Four Deployment Models

The four models highlight the differences in how organizations handle civilian deployments. We believe that organizations can draw on a combination of these models, emphasizing aspects that fit their specific situation and best position them to achieve their overall objectives. For example, it was clear that most organizations with a requirement to deploy civilians on relatively short notice to a hostile environment chose to develop a kind of cadre; they had a process for preselecting people who could fill requirements that arose quickly.[56] However, we found that the time it took to fill a requirement for a particular civilian deployment varied greatly. A number of factors affected the speed of deployment, including the organization from which the individuals were sourced, the extent to which the required skill set was readily available, and the generic selection procedures applied within the organization.

Yet, we found it notable that, of the four models we identified, those that involved a proactive sourcing approach allowed organizations to deploy personnel significantly faster than those that involved

[56] Interview with a FEMA official, September 19, 2014; interview with a CS3 official, September 15, 2014; interviews with S2O officials, August 2014; interview with a DSTO official, September 25, 2014.

recruiting personnel after a requirement had been issued.[57] Meanwhile, organizations that relied on external sourcing models spoke to the numerous rules and regulations faced by government agencies. At times, lengthy justifications were needed to select one individual over another to adhere to fair hiring practices.[58] This indicates that, when reactive sourcing is necessary, sourcing officials may need direct or expedited hiring authorities to enhance their ability to source positions quickly. Furthermore, regardless of whether individuals were sourced internally or externally, some type of oversight organization was necessary to ensure the successful deployment of civilians.

Across the organizations analyzed, we identified opportunities to pool and share existing capabilities for civilian deployments. For instance, certain agencies within the Intelligence Community share predeployment training and medical facilities, such as DIA's Expeditionary Readiness Center.[59] Instead of maintaining individual deployment divisions, organizations could pool those resources, and one agency could provide training on behalf of the others.

To source and deploy civilians rapidly, our analysis suggests the need to closely examine the speed of the recruitment process, the possibility of developing a preselected pool, and the possibility of making deployment part of the job description.

Finally, there are a number of decisions to be made with regard to the size and scope of a deployable civilian capability. Planning and forecasting will help optimize the timelines associated with deployment. For instance, if the requirement is not urgent, the organization has time to use a reactive sourcing model. Although we did not directly assess the difference in costs between sourcing external candidates versus internal candidates, it is likely that costs will differ and that the

---

[57] Experts from the CPCC Civilian Response Team pool have been deployed within five days. Interview with a CPCC official, October 10, 2014; interviews with U.S. government officials, 2014.

[58] Interview with a CPCC official, October 10, 2014; interview with an APSPO official, September 10, 2014.

[59] Interviews with DIA officials, September 2014.

cost itself will be a factor when choosing a sourcing model.[60] Furthermore, if a skill set is required within the organization in the future, it is valuable to maintain and sustain the skill set internally rather than externally. Establishing any type of preidentified readiness pool will therefore necessitate the accurate forecasting of future requirements and likely mission sets.

Related to this point, the organizations analyzed considered the positions that they were looking to fill with civilians; they also scrutinized the requirements for civilian deployment to ensure that the post was necessary and that only a civilian could fill it.[61] For instance, the Canadian Department of National Defence, Australian DSTO, UK MOD, and EEAS CPCC all draw their civilian deployees from volunteers—that is, individuals deploy on a voluntary basis. Interviewees from these organizations noted that, within their workforces, they had capable individuals who were interested in volunteering, and they found that very few people withdrew their offer to deploy.[62]

Meanwhile, for the U.S. agencies examined here, some mandated that specific individuals deploy (Bureau of Diplomatic Security, Bureau of Conflict and Stabilization, OTI, DLA),[63] others requested volunteers (the U.S. Department of Homeland Security's Office of International Affairs, CBP, DEA),[64] and still others used a combination of factors to make this decision, including the source of the requirement, the time needed to fill the position, and whether the skill set was internal to the organization. Organizations within the DoD Fourth Estate typically had a mixed civilian/military workforce. In organizations with a well-

---

[60] Such a comparative cost analysis assessing the relative expense of each of the four deployment models outlined here would be a fruitful area for future research.

[61] Interviews with S2O officials, August 2014; interviews with Canadian Department of National Defence officials, September 25, 2014.

[62] Interviews with officials from analogous organizations, May–November, 2014.

[63] Interview with a Bureau of Diplomatic Security official, August 14, 2014; interview with a Bureau of Conflict and Stabilization Operations official, September 18, 2014; interview with an OTI official, September 15, 2014; interview with a DLA official, July 19, 2014.

[64] Interview with an Office of International Affairs official, August 25, 2014; interview with a CBP official, September 18, 2014; interview with a DEA official, September 24, 2014.

defined set of requirements to deploy personnel, there were usually systems in place to facilitate individuals volunteering to fill a requirement.

In sum, most of the organizations examined had well-defined policies that clearly articulated duties and procedures surrounding the deployment process. However, other agencies lacked many basic policy documents and consistently handled their deployment procedures on an ad hoc basis.[65]

If the permanent DoD civilian workforce is designed to include all necessary expeditionary civilian capabilities internally, the nature of the workforce may change. In return, that skill set will remain within DoD once the deployment is completed. For some skills, this may be critical if a future surge is required. Other skills, however, may not need to be retained internally (for example, Ebola/infectious disease specialist physicians). Yet, overutilization of expeditionary civilian personnel sourced from outside DoD will pose challenges to postdeployment tracking, and the skill sets will not be readily available in the future. Such considerations speak to a need to forecast future requirements effectively and to appropriately balance external versus internal sourcing and proactive versus reactive sourcing in light of both these forecasts and projected time frames for deployment. The next chapter explores the operationalization of such measures in greater detail.

---

[65] Interviews with officials from various non-DoD organizations, May–November, 2014.

CHAPTER FIVE

Conclusion and Recommendations: A Framework for the Future

The primary goals of this study were to (1) gauge whether CPP is on the right track with its current DoD civilian deployment model, and (2) to recommend guidelines for establishing and maintaining a civilian deployment capability that meets CCMD requirements. In this chapter, we present conclusions related to the first goal and recommendations in line with the second.

Conclusions

Overall, we found that the current practice of DoD civilian deployment requires certain modifications to be viable over the long term. We reached this conclusion for several reasons. First, the current sourcing model contains elements that existed prior to DoDD 1404.10—namely, decentralized recruitment, screening, and selection of civilian candidates *by the individual military services* for deployable position requirements. These sourcing practices did not work in the past, in the absence of the policy articulated in DoDD 1404.10. As noted in Chapter One and elsewhere in the report, this was reflected in multiple interviews. DoD officials highlighted the Army's difficulty filling expeditionary civilian positions in the 2007–2009 time frame because civilians were not adequately incentivized to deploy, and the services that owned them were given little incentive to support their deployments.[1]

[1] Interviews with DoD, CCMD, and U.S. military officials, May–November 2014.

Second, the sourcing model employed by USCENTCOM—with a separate CEW office housed at the CCMD that is responsible for sourcing (recruitment, screening, and selection of civilian candidates)—is not entirely applicable to other CCMDs. Therefore, this model is not broadly generalizable for potential future missions.

Third, the extent to which military planning actually considers expeditionary civilian requirements is questionable at this point, at least as indicated by our interviews. A failure to effectively integrate expeditionary civilians into planning for various scenarios and missions hinders the development of realistic expectations for the number of expeditionary civilians required and ultimately poses a challenge to the services by decreasing their ability to plan for backfill needs when one of their civilians deploys.

Fourth, many CCMDs appear unlikely to utilize expeditionary civilian capabilities in the future, as they are unaware of the benefits that expeditionary civilians can provide or feel that they have no need for additional expeditionary civilian capabilities beyond their own current capacity. This could change, however, as their mission sets evolve or if they become more familiar with expeditionary civilian capabilities.

Finally, our interviews uncovered conflicting perceptions across DoD regarding the costs of deploying DoD civilians versus U.S. military personnel. Such conflicting perceptions are potentially problematic for the long-term viability of DoD civilian deployment, because perceptions of cost could drive organizations to either push for or hamper the deployment of DoD civilians. Moreover, it is difficult to define the roles of DoD expeditionary civilians with respect to the various components of the Total Force when the relative costs of deploying them remain a mystery.

Recommendations

To devise guidelines for strengthening the long-term viability of the practice of civilian deployment, we identified three pertinent issues for consideration:

1. What are the relevant aspects of a civilian deployment capability that must be maintained over the long term?
2. Who has responsibility for these various aspects of a civilian deployment capability?
3. What tasks should responsible parties undertake to ensure the establishment and long-term maintenance of a civilian deployment capability to meet CCMD requirements?

We consider each of these issues in the following sections and recommend future actions and activities to ensure the long-term viability of civilian deployments, where appropriate.

"Ownership" of Relevant Aspects of Civilian Deployment Capabilities

It is useful to consider the management of DoD civilian deployment capabilities as being divided into three categories of activities: policy, planning and strategy, and operations. *Policy* responsibilities entail identifying guidelines for civilian deployment. *Planning and strategy* responsibilities entail mission-based, scenario-specific forecasting and SHCP. *Operational* responsibilities entail assigning requirements, sourcing, readiness preparation, and tracking expeditionary civilians during and after deployments. Figure 5.1 shows these distinctions.

Policy Responsibilities

Policy responsibilities for DoD civilian deployment clearly fall within the purview of OUSD(P&R), and specifically CPP—a policy office whose core competency involves writing policies on DoD civilian workforce issues. Two of our recommendations would most likely fall within the realm of the policy owner for expeditionary civilian capabilities: raising awareness of expeditionary civilian capabilities and performing a comparative cost study to inform Total Force planning.

Champion Expeditionary Civilian Capabilities

As noted in Chapter Three, our interviewees indicated that many CCMDs are unaware of the potential benefits that expeditionary civilians can provide and require better knowledge up front of the potential range of these capabilities. This could be explained by the relatively

Figure 5.1
Ownership of Relevant Aspects of Civilian Deployment

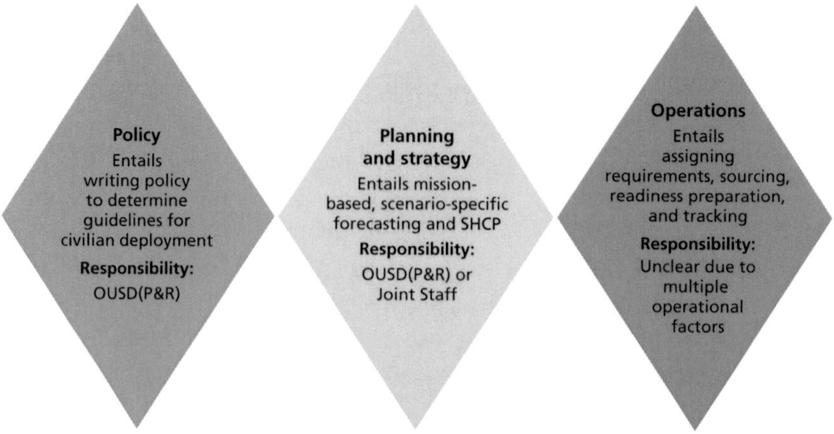

RAND *RR975-5.1*

few resources that were put into raising awareness of civilian capabilities across the U.S. military. Interestingly, we found that all of the analogous organizations examined in our comparative analysis in Chapter Four marketed their deployable civilian capabilities to some extent to raise awareness of these capabilities and the implications of deployments.[2] For these organizations, raising awareness serves two primary functions. First, it provides a realistic picture of what it means to be a deployed civilian. This gives applicants a realistic preview of the positions in which they could possibly deploy and helps adjust expectations for both those generating requirements and the civilians who consider volunteering for a deployment. Second, raising awareness ensures that individuals who issue requests for deployed civilians understand both the current and future capabilities of this workforce. This helps those generating requirements understand where civilian deployees would be most appropriate and how they could be employed.

---

[2] Interviews with DIA officials, October 10, 2014; interviews with CS3 and OTI officials, September 2014; interview with an APSPO official, September 10, 2014; interview with a DEA official, September 24, 2014; interview with a CBP official, September 18, 2014.

To raise awareness in the military domain, there are several ways to effectively market civilian contributions to contingency operations. One option is to engage military commanders who have experience commanding expeditionary civilians to talk about the civilians' effects on the mission. Understanding the experience of civilians before and after deployment is also important and can lead to more frequent utilization of the civilian workforce.

One organization in our study offered a second option: having senior defense civilians promote civilian deployment capabilities. On numerous occasions, senior civilians worked with their internal and external media relations teams to increase the visibility of the civilian component of operations across the workforce. One example of this was the use of defense blogs and articles describing the experiences of civilian deployees.[3] Analogous organizations are also working toward more public recognition of deployed civilian contributions by rewarding deployees with medals and other awards.[4]

All of these initiatives help generate an attractive career option for civilians and educate potential customers of expeditionary civilian capabilities, as well as the rest of the defense workforce, about the potential operational value of civilian deployees. Such initiatives thus support a cultural shift in overcoming potential misconceptions about the availability, deployment process, and usability of civilians in support of military operations. *Because our research highlights a need for improved awareness to ensure a future demand for expeditionary civilian capabilities across the CCMDs, we recommend that DoD policy owners undertake efforts to raise awareness of these capabilities.* This responsibility might also feasibly fall to the DoD entity responsible for planning and strategy, as raising the awareness of these capabilities could be construed as relevant to either policy or planning.

---

3 See, for instance, UK Ministry of Defence, "Preparing Civilian Staff for Operational Deployments," September 14, 2012; Defence Science and Technology Laboratory, "DSTL Support to Operations," RSS channel, last updated July 28, 2014.

4 Interviews with MOD officials, August 2014.

Perform a Study Comparing the Relative Costs of Deploying DoD Civilian, Military, and Contractor Personnel

As discussed in Chapter Two, there is a dearth of analysis comparing the relative costs of deploying civilians, contractors, and military personnel. Because such information is critical to the development of a cost-efficient Total Force policy specifying distinct roles for civilians, contractors, and military personnel deployed to a theater of conflict, *we recommend that policy owners conduct or oversee a comparative cost study along these lines.* The results from this analysis could then inform force-sizing initiatives within the services, CCMDs and the Joint Staff. We suggest a three-step method for such a study, to capture numerous variables simultaneously in a robust manner:

1. Develop a comprehensive cost structure considering the security costs for various personnel types, as well as salary and associated benefits.
2. Create a data set of representative military, civilian, and contractor personnel characteristics to control for experience and other skills or attributes simultaneously with cost (to determine whether more expensive personnel are actually more effective and thus worth the price).
3. Consider scenarios in which each group would be most cost-effective.

Planning and Strategy Responsibilities

Responsibility for planning and strategy related to expeditionary civilians is slightly less clear, though existing policy guidance points to a lead role for OUSD(P&R) as well, in terms of its SHCP responsibilities. As discussed in Chapter Two, DoDI 1400.25, Volume 250, states,

> A structured competency-based approach will be instituted throughout the Department of Defense in support of SHCP that applies job analysis methodologies compliant with the Uniform Guidelines on Employee Selection Procedures . . . and meets the requirements and objectives of References (c), (d), and (e). This approach will be used to identify current and future civilian workforce requirements, *including those of an expeditionary nature*, as

> part of total force planning. It will also be used to establish a plan to ensure the readiness of the civilian workforce to meet those requirements. Control of resources, management, and execution will remain with the DoD Components in the implementation of this policy. [Emphasis added][5]

This policy places responsibility for leading SHCP efforts with the Office of the Deputy Under Secretary of Defense for Civilian Personnel Policy and the Defense Human Capital Strategy Program Executive Office (DHCS PEO). Specific responsibilities for SHCP with regard to expeditionary civilians also fall to these offices, according to this guidance:

> The DUSD(CPP) and the DHCS PEO, for the duration of that office, both under the authority, direction, and control of the USD(P&R), working with the DoD Components, will co-lead the effort to develop and establish a cohesive DoD-wide direction for human capital, including competency-based SHCP to ensure the readiness of the civilian workforce to meet mission requirements. SHCP implementation guidance will be developed jointly with the DoD Components, the functional communities, the DUSD(CPP), and the DHCS PEO. Through a dedicated Program Office (PO), they shall provide guidance to Functional Community Managers (FCMs) in the execution of this Volume, and shall . . . Monitor the strategic environment, workforce trends, competency assessments, and gap analyses to ensure that recruitment, retention, and development initiatives address DoD current and future mission requirements, *including those of an expeditionary nature* . . . [and] Oversee the compilation of civilian workforce and capability requirements, including those that may require expeditionary deployments. [Emphasis added][6]

Since this guidance was issued in 2008, CPP has transitioned from a deputy under secretary of defense–level office to a deputy assistant secretary of defense–level office. It is unclear whether this alters

---

5 DoDI 1400.25, Vol. 250, 2008, p. 3.

6 DoDI 1400.25, Vol. 250, 2008, Enclosure, p. 4.

the extent of its responsibility for the planning and strategy aspects of DoD civilian deployment, but the possibility is worth noting.

One of our recommendations would most likely fall within the realm of the planning and strategy owner for expeditionary civilian capabilities: an increase in mission-specific, scenario-driven planning and forecasting of future requirements for expeditionary civilians.

Perform Mission-Specific, Scenario-Driven Forecasting

As noted in Chapter Two, GAO reported in 2012 that

> DoD components have not identified and designated the number and types of positions that should constitute the CEW because guidance for making such determinations has not been provided by the Office of the Secretary of Defense. . . . Until guidance that instructs the components on how to identify and designate the number and types of positions that will constitute the CEW is developed, DoD may not be able to (1) *make the CEW a significant portion of the civilian workforce as called for in DoD's fiscal year 2009 Civilian Human Capital Strategic Plan,* (2) *meet readiness goals for the CEW as required in DoD's Strategic Management Plan for fiscal years 2012–2013*, and (3) position itself to respond to future missions. [Emphasis added][7]

We therefore strongly recommend more extensive manpower planning and forecasting pertaining to future requirements for DoD expeditionary civilians. Such planning is needed to delineate appropriate numbers of E-E–coded personnel and enact E-E coding to a greater extent and to help reduce uncertainty regarding the numbers of service and Fourth Estate civilians who may have to deploy in a given contingency (see Figure 5.2).

As discussed in Chapters Three and Four, our comparative case analysis and interviews indicated that utilization will occur only if the capability is viewed as reliable, the capability will be viewed as reliable only if it is "ready," and expeditionary civilians will be ready to deploy on short notice only if they are coded as E-Es and proactively sourced

---

[7] GAO, 2012.

Figure 5.2
The Benefits of Scenario-Based, Mission-Specific Forecasting

RAND *RR975-5.2*

to at least some extent. The forecasting recommended here is therefore necessary to ensure CCMD utilization of expeditionary civilian capabilities, to inform the development of a tiered sourcing model and to support broader Total Force management efforts over the long term. At least 15 people we spoke with across the services, Joint Staff, Fourth Estate, and OSD—including multiple senior-level OSD and military officials—thought that mission-specific, historically based modeling would be useful in designing a viable long-term civilian deployment practice.[8]

Such forecasting should ideally take into account the range of plausible scenarios or hypothetical contingencies (e.g., missions, locations, duration). It should also incorporate a force-sizing construct that provides the strategic rationale for a particular force posture, including

[8] Interviews with military, Joint Staff, Fourth Estate, and OSD officials, May–November 2014.

roles and responsibilities. Relevant assessments might emerge from the various DoD contingency-planning efforts conducted under the aegis of the Guidance for Employment of the Force and the Joint Strategic Capabilities Plan, as well as in DoD's Support to Strategic Analysis process. However, for this to be successful, such planning would have to estimate demand for nonstandard capabilities and expeditionary civilians not captured during the formal assignment of expeditionary civilian requirements under the GFM process.

The demand signal that the IES Office was launching at the time of this writing is a significant and commendable step in the right direction with regard to this recommendation. *We recommend building on this demand signal work to incorporate different mission sets into the planning and to look more deeply into historical requirements for deployed civilians across scenarios and missions.* Moreover, as mentioned in Chapter Two, CPP plans to turn over demand signal maintenance responsibilities to the services; however, the services lack an enterprise-wide view of expeditionary civilian requirements and available capabilities, indicating a need for some degree of joint oversight of this responsibility. *We therefore recommend that responsibility for maintaining and updating the demand signal remain within OUSD(P&R).* It is critical to note that if this responsibility remains with CPP, the office will likely need additional resources and manpower to effectively fulfill this role.

Operational Responsibilities

Ownership of operational responsibilities for DoD civilian deployment is more complicated than these other areas of management responsibility. This is because of the complex, multifaceted nature of the operations for which civilians deploy. As noted earlier, operational responsibilities entail the assignment of requirements, sourcing, readiness preparation, and tracking.

The first step in the operational planning for civilian deployment involves the formal *assignment of expeditionary civilian requirements* to the various DoD components in the Secretary of Defense Operations Book. As discussed in Chapter Two, civilians are not provided with career-based incentives to deploy, and their home offices are not motivated to support their deployment because of the lack of backfill. Moreover, the military services and Fourth Estate agencies lack a real-time view of available expeditionary civilian capabilities across the DoD civilian workforce. It therefore appears that some type of joint oversight is needed to supervise the assignment of requirements, to compel home offices to let their civilians deploy, and to provide career-based incentives for civilians to deploy, as well as to find the correct skill sets from across the services, Fourth Estate, and external sources, if needed.

As discussed in Chapter Three, *sourcing* involves recruiting, screening, and selecting candidates, as well as providing the candidates. Because sourcing processes run the risk of becoming stovepiped within organizations with different criteria for screening, vetting, and selecting candidates, some type of joint oversight would be useful to ensure a balanced workforce in which expeditionary civilians are sourced similarly across the services and Fourth Estate. Such joint oversight of recruitment, screening, and selection would also be useful in seamlessly integrating any *external* candidates into the overall DoD sourcing process. Because the military services and DoD Fourth Estate agencies own any DoD civilians who might deploy from the internal labor pool, the second aspect of sourcing—at least in terms of the provision of *internal* candidates—should be their responsibility.

Readiness preparation involves training candidates, processing any necessary clearances (e.g., medical, security), and providing visas, passports, and other administrative documentation necessary for deployment. While the services are best positioned to manage the readiness preparation of DoD civilian employees tasked with service missions, they may not be able to manage readiness preparation for civilians who are deploying to joint commands to conduct joint missions. A separate readiness process may be necessary for civilians hailing from either the Fourth Estate agencies or sources outside of DoD to ensure the future viability of DoD civilian deployment.

Finally, *tracking* entails maintaining contact with a civilian deployee both during deployment and for a period of time postdeployment, to assist with administrative, human resources, or occupation-related issues and to screen the individual for any deployment-related health problems. As explained in more detail later, we recommend that a nonservice, joint-level organization be tasked with oversight and management of the assignment of requirements for expeditionary civilians, as well as recruitment, screening, and selection. The same joint-level organization should be tasked with tracking civilian deployees both during and after deployment to ensure the comprehensiveness and uniformity of employer-provided support and services.

Provide or Arrange for Joint-Level Oversight of Operations

As indicated earlier, we identified several key operational responsibilities as being suited to a joint-level organization: assignment of requirements and sourcing, tracking, and readiness preparation of DoD civilians who are not deploying to support service missions. While we make no claim as to the best joint-level "owner" of these functions, we explored several potential alternative models for joint-level ownership of DoD civilian deployment operations and highlight the pros and cons of each. This is not an exhaustive list of the possibilities, but it provides some idea of the range of options available to DoD, as well as the potential benefits and drawbacks of each option.

The first option we explored is the CCMD sourcing model, which would involve a separate CEW office or point of contact at each of the CCMDs tasked with sourcing expeditionary civilian requirements. This was the model being established at USCENTCOM as of the time of this writing. There are several potential benefits associated with the CCMD sourcing model, including knowledge among CCMD personnel of the intricacies of the position requirements when screening resumes.[9] Moreover, because this is a model not housed within any one service, it has joint characteristics with the potential to provide an enterprise-wide view of available capabilities and to rise above the incentive problems associated with decentralized sourcing.[10]

However, there are drawbacks associated with the CCMD sourcing model as well, such as USCENTCOM's uniqueness as the only CCMD with both the need and resources for this type of model; it does not appear to be generalizable across the other CCMDs, which currently have substantially lower demand for deployable civilian capa-

---

[9] Interview with a USCENTCOM official, August 28, 2014.

[10] Correspondence with a USCENTCOM official, September 3, 2014.

bilities.[11] Moreover, because CCMD staffs are mainly military, officials have argued that they are not well skilled in the intricacies of civilian hiring.[12] Finally, the CCMD sourcing model does not meet the intent of a deployable cadre as outlined in the 2007 NDAA and DoDD 1404.10, as it is purely reactive.

The second option we explored was OUSD(P&R) or Joint Staff ownership of the assignment of expeditionary civilian requirements, as well as recruitment, screening, selection, tracking, and readiness preparation and training for non-service (i.e., Fourth Estate) civilians.[13] Joint-level oversight of these operational tasks would act as a forcing function to realign both service and civilian incentives for civilian deployment, because such a joint-capable organization could oversee and compel the services to provide civilian capabilities to meet requirements. Joint oversight could also enforce appropriate treatment of deployees by their home offices upon their return. Oversight of sourcing is already technically a Joint Staff responsibility,[14] and the readiness of CEW personnel has historically been managed under the auspices of OUSD(P&R), so ownership of these operational functions for civilian deployment by either entity would be credible and appropriate in light of their respective missions. Finally, having such high-level joint oversight of civilian deployment operations would help ensure that expeditionary civilian capabilities are structured in the context of Total Force management.

---

[11] Interview with a USPACOM official, September 22, 2014; interview with a USEUCOM official, September 22, 2014; interviews with USAFRICOM officials, September 30, 2014; interview with a USNORTHCOM official, November 5, 2014.

[12] Correspondence with a DoD official, October 29, 2014.

[13] While this option would entail joint oversight of the assignment of requirements, recruitment, screening, selection, and tracking for all expeditionary civilians, it would also entail joint oversight of readiness preparation and training *only* for civilians not "owned" by a service, such as those employed by Fourth Estate agencies. This is because the readiness preparation and training for civilians employed by the military services would be handled directly by the services.

[14] Chairman of the Joint Chiefs of Staff Instruction 1301.01F, 2014, Enclosure B; interview with an Air Force official, July 21, 2014; interviews with DoD officials, May 30, 2014.

However, the Joint Staff may not have the capacity to take on this responsibility.[15] Moreover, OUSD(P&R) involvement would have to be managed carefully to be seen as legitimately within the office's mandate. OSD-level offices are not typically involved in operational efforts, holding primary responsibility for creating policy. However, several interviewees believed that *oversight* of civilian deployment operations would appropriately fall under OUSD(P&R).[16] If either OUSD(P&R) or the Joint Staff were to take on ownership of the operational processes described here, these processes would have to be resourced sufficiently. Despite these challenges, the substantial benefits associated with this option make it relatively attractive.

The third option we explored would involve designating one Fourth Estate agency (for example, DLA) to act as the executive agent for *all* expeditionary civilian capabilities. This executive agent would be responsible for recruitment, screening, selection, readiness preparation, training, and tracking of all DoD expeditionary civilians. A benefit of this approach is that it would provide some degree of joint oversight and visibility across the services by a nonservice entity. Therefore, it could act as a forcing function to realign service and civilian incentives for civilian deployment. Another notable benefit of this model is that several Fourth Estate agencies already have well-tested and successful deployment models, as well as existing readiness centers, that might feasibly be adopted and used for all DoD expeditionary civilians.[17] Nonetheless, a question remains regarding how to choose a particular Fourth Estate agency for this responsibility and how to ensure that it is resourced appropriately for this function. Moreover, there is nothing to guarantee that the Fourth Estate agency selected would have adequate incentives to source positions from across the other DoD components and not privilege its own personnel for certain positions, leaving open

[15] Interviews with DoD officials, May–November 2014.

[16] Interviews with DoD officials, October 16, 2014; interview with a former senior DoD official, October 20, 2014.

[17] For example, as noted in Chapter Four, DIA's Expeditionary Readiness Center provides predeployment training for both DIA employees and deployees from a variety of other organizations. Interview with a DIA official, November 14, 2014.

the possibility that existing shortcomings in the deployment of expeditionary civilians would remain.

The final model that we considered for joint-level oversight of operations is the designation of distinct components or agencies to act as executive agents for different expeditionary civilian mission sets, with each executive agent being responsible for recruitment, screening, selecting, training, and tracking civilians within its respective mission set. A benefit of such a model is that, as with all of those outlined previously, joint-level oversight would force the realignment of service and civilian incentives for civilian deployment. Moreover, this model could benefit from the well-tested deployment models and existing readiness centers of various Fourth Estate agencies. The unique appeal of this model derives from the better positioning of certain DoD components and agencies to operate with particular mission sets. For instance, the U.S. Army Corps of Engineers is well placed to perform missions requiring extensive engineering expertise, such as stabilization and reconstruction.

However, numerous drawbacks are associated with this model, including the challenge of determining which components or agencies should be assigned operational responsibilities for civilian deployment in particular missions, and, relatedly, a risk of turf battles between executive agents when certain operations include a combination of missions. Finally, the risk remains under this model that expeditionary civilians will not be treated uniformly in terms of preparation, tracking, and performance reviews, among other human resources–related matters.

Adopt a Tiered Sourcing Approach

As noted earlier, our comparative case analysis and interviews indicated that CCMD utilization of expeditionary civilian capabilities will occur only if the capability is viewed as reliable. In turn, our interviews indicated that expeditionary civilian capabilities will be viewed as reliable only if individuals are able to deploy relatively quickly.[18] Moreover, our

---

[18] Interview with a USNORTHCOM official, November 5, 2014; interview with a USCENTCOM official, August 28, 2014; interview with an Air Force official, July 21, 2014.

comparative analysis of analogous cases indicated that expeditionary civilians will likely be ready to deploy on short notice only if they are E-E–coded and proactively sourced *to at least some extent*. This does *not* necessarily have to be a large-scale standing cadre as originally envisioned for the CEW, however; it could be a "tiered approach" involving different categories of personnel.

Based on our analysis of analogous organizations, we recommend adopting a tiered sourcing approach for expeditionary civilian capabilities. The foregoing analysis indicates that a tiered sourcing approach is suited to organizations that have a need to deploy a wide range of civilians within differentiated time frames. We found that organizations with quick-turn requirements are best served by a proactive sourcing approach. For positions with less urgent deployment requirements or for which civilian skills must meet an emerging requirement, organizations should consider a reactive sourcing approach. The tiered approach combines these two models and will allow an organization to effectively meet both short-notice, emerging requirements and planned, enduring requirements in a timely and cost-effective manner.

Figure 5.3 depicts a tiered sourcing model. At the top is the readiness pool, which consists of personnel ready to deploy on short notice. Personnel in this pool could be sourced internally (either E-E coded or volunteers) or externally and could be full-time or at-will employees. Both the bureau or office in charge of the readiness pool and the readiness pool itself must be managed carefully, such that the readiness of individuals is tracked and all qualifications for the readiness pool are met (e.g., medical screening, training, security clearance, passports). As challenges arise with personnel, they should be met directly. It is critical to keep morale high within this group so that the pool does not suffer from unmet expectations. Moreover, it is critical to regularly review and revalidate all formal and informal position requirements and personnel in the readiness pool to ensure that the appropriate skill sets are available to meet evolving requirements.

The second tier of the model comprises E-E–coded individuals. Personnel at this level will be in positions coded in a way that will force them to deploy if a requirement arises that fits their particular skill set. Notably, these personnel will not be kept at the same level of readiness

Figure 5.3
Tiered Sourcing Model

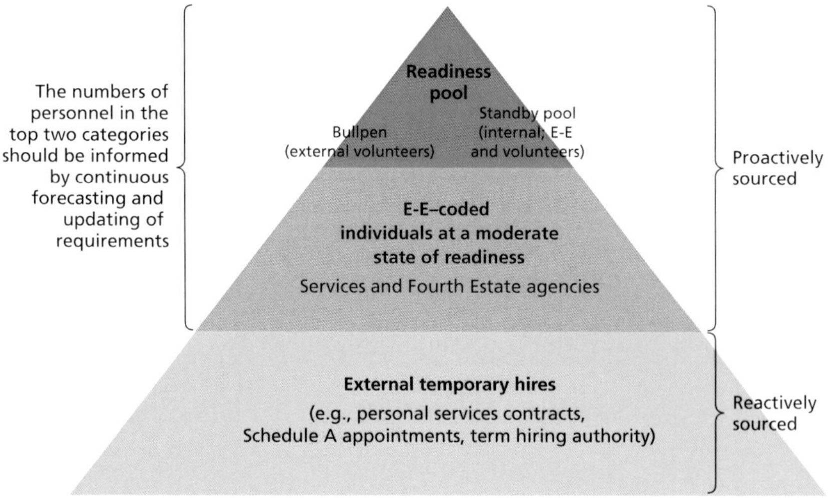

RAND *RR975-5.3*

as those in the readiness pool. For example, they could have completed some predeployment training and have a passport and security clearance, but they may not have completed all necessary medical screening in advance. Individuals at this level will have more lead time to accomplish predeployment tasks.

One concern regarding a pool of deployable civilians is that when they are called to deploy, there is no forcing function in place. It is therefore critical for the organization to be systematic in utilizing civilians for their expertise and carefully scoping the likely requirements that E-E–coded individuals will be called to fill. When hired, these personnel need to understand that their position codes them in a manner that could force them to deploy.

The last tier of the model consists of external temporary hires. These individuals will be hired from outside the organization and used to fill requirements only once the top two tiers have been fully utilized. Individuals in this tier could be hired under a variety of mechanisms, but upon completing their tours, they will not be entitled to jobs in the home organization. These individuals will be term employees and hired only when surge requirements arise that outstrip capacity in the top two tiers. It is important to note that the requirements for individuals in all three tiers of the model should be informed by the mission-specific, scenario-based forecasting recommended earlier and should continually be updated on a regular basis to coincide with emerging needs and operational demands.

In sum, it appears that several modifications to the existing practice of DoD civilian deployment are necessary to ensure the viability and utilization of DoD expeditionary civilian capabilities over the long term. However, we do not expect any of these recommendations to be overly burdensome. In fact, if implemented correctly, they could improve efficiency and cost savings across DoD as it seeks to meet its operational missions through improved Total Force planning and strategic human capital management.

APPENDIX A

Glossary of Key Terms

Bullpen: A readiness pool consisting solely of externally sourced individuals.

Cadre: A collection of people designated for a particular purpose, whether or not it is managed as a career pathway, a functional/occupational group, or just a group of people.

Decentralized sourcing by services/executive agents: A sourcing model whereby the military services and CCMD executive agents are responsible for filling requirements, preparing civilians for deployment, ensuring that civilian deployees are supported while deployed, and ensuring that civilian deployees are able to return to their jobs—or equivalent positions in their home organizations—upon their return.

External sourcing: The practice of hiring for expeditionary positions outside the existing DoD workforce. External sourcing typically entails the use of temporary hiring authorities.

External temporary hiring: Employing individuals from outside DoD through a variety of term-employment mechanisms. Upon completing their tour, these personnel will not be entitled to a job within the home organization.

Internal sourcing: The practice of hiring for expeditionary positions within the existing DoD workforce—from among the military services, Fourth Estate agencies, OSD, Joint Staff, and similar entities.

Joint civilian requirements: Requirements that are *intended to fill joint missions*, regardless of whether they are issued through a JIA request or RFF.

Joint oversight of operations: Oversight of a non-service entity over deployment, recruitment, training, and tracking of expeditionary civilians during and after deployment.

Planned deployments: Anticipated deployments included in an organization's long-term planning and for which the organization may therefore feasibly preidentify individuals for the requisite positions.

Proactive sourcing: Preidentifying individuals to fill potential positions prior to the issuance of requirements.

Reactive sourcing: Identifying individuals to fill positions after requirements have been issued.

Readiness pool: A preidentified group of personnel ready to deploy on short notice. Personnel in this pool could be sourced internally or externally and could be full-time or at-will employees. The readiness of individuals in this pool is continually tracked and updated. It is critical to regularly review and revalidate all formal and informal position requirements and personnel in the readiness pool to ensure that the appropriate skill sets are available to meet evolving requirements.

Service civilian requirements: Requirements that are *intended to fill a service-specific function.*

Short-notice deployments: Quick-turn requirements that may or may not be anticipated asking for an individual to be ready to deploy within a matter of weeks, rather than a month or more.

APPENDIX B

Organizational Affiliations of Interviewees

Australian Defence Science and Technology Organisation (now the Defence Science and Technology Group)

Canadian Department of National Defence

Defense Contract Management Agency

Defense Intelligence Agency

Defense Logistics Agency

European External Action Service, Civilian Planning and Conduct Capability

Joint Staff, J1 (Manpower and Personnel)

Joint Staff, J35 (Future Operations)

Joint Staff, J4 (Logistics)

Joint Staff, J5 (Strategic Plans and Policy)

Office of the Deputy Assistant Secretary of Defense for Civilian Personnel Policy

Office of the Deputy Assistant Secretary of Defense for Program Support

Office of the Deputy Assistant Secretary of Defense for Readiness

Office of the Under Secretary of Defense for Intelligence

Office of the Under Secretary of Defense for Personnel and Readiness, Office of Total Force Planning and Requirements

UK Ministry of Defence, Support-to-Operations Team

U.S. Africa Command

U.S. Agency for International Development, Office of Transition Initiatives

U.S. Agency for International Development, Crisis Surge Support Staff

U.S. Air Force

U.S. Army

U.S. Central Command

U.S. Department of Homeland Security, U.S. Customs and Border Protection

U.S. Department of Homeland Security, Federal Emergency Management Agency

U.S. Department of Homeland Security, Office of International Affairs

U.S. Department of Justice, Drug Enforcement Administration

U.S. Department of State, Afghanistan and Pakistan Strategic Partnership Office

U.S. Department of State, Bureau of Conflict and Stabilization Operations

U.S. Department of State, Bureau of Diplomatic Security

U.S. European Command

U.S. Navy

U.S. Northern Command

U.S. Pacific Command

U.S. Southern Command

U.S. Special Operations Command

Bibliography

Acquisition Advisory Panel, *Report of the Acquisition Advisory Panel to the Office of Federal Procurement Policy and the United States Congress*, Washington, D.C., January 2007.

Army National Guard, "Constitutional Charter of the Guard," 2009.

Association of the U.S. Army, "Army Total Force Policy: Fully Integrating the Operational Reserve," *Defense Report*, Arlington, Va., December 2012. As of April 2, 2016:
http://www.ausa.org/publications/ilw/ilw_pubs/defensereports/Documents/DR_12-1_web.pdf

Bensahel, Nora, and Patrick M. Cronin, *America's Civilian Operations Abroad: Understanding Past and Future Requirements*, Washington, D.C.: Center for a New American Security, January 2012.

Berteau, David, Joachim Hofbauer, Jesse Elman, Gregory Kiley, and Guy Ben-Ari, *DoD Workforce Cost Realism Assessment*, Washington, D.C.: Center for Strategic and International Studies, May 2011.

Binnendijk, Hans, and Patrick M. Cronin, eds., *Civilian Surge: Key to Complex Operations*, Washington, D.C.: National Defense University Press, 2009.

Blanks, Julie, International/Expeditionary Support Office, "CEW Transition," briefing, May 3, 2013.

———, "Deployable Civilian Demand Signal," briefing, October 2014.

Carty, Kevin, *Good Practices in Basic Police Training—Curricula Aspects*, Vienna, Austria: Organization for Security and Co-operation in Europe, May 2009.

Chairman of the Joint Chiefs of Staff Instruction 1001.01B, *Joint Manpower and Personnel Program*, October 7, 2014.

Chairman of the Joint Chiefs of Staff Instruction 1301.01F, *Joint Individual Augmentation Procedures*, November 17, 2014.

Chlosta, Matthew, "'AfPak Hands' Begin Immersion Training," American Forces Press Service, May 5, 2010. As of April 2, 2016:
http://archive.defense.gov/news/newsarticle.aspx?id=59031

Chu, David S. C., Under Secretary of Defense for Personnel and Readiness, "Interim Policy and Procedures for Strategic Manpower Planning and Development of Manpower Estimates," memorandum, December 10, 2003.

———, "Building Increased Civilian Deployment Capacity," memorandum, February 12, 2008.

Civilian Personnel Policy/Defense Civilian Personnel Advisory Service, Civilian Expeditionary Workforce Program Office, "CEW Support Costs and FTE Breakdown Version 3," provided to RAND in April 2014.

Cole, Ronald, "Grenada, Panama, and Haiti: Joint Operational Reform," *Joint Force Quarterly*, Fall–Winter 1998–1999.

Commission on Wartime Contracting in Iraq and Afghanistan, *Transforming Wartime Contracting: Controlling Costs, Reducing Risks*, Arlington, Va., August 2011.

Corporation for National and Community Service, "FEMA Corps," web page, undated. As of April 2, 2016:
http://www.nationalservice.gov/programs/americorps/fema-corps

Cox, J. David, Sr., "DoD Insourcing Makes Financial Sense," *Federal Times*, October 29, 2013.

Cronin, Patrick M., *The Total Force Policy in Historical Perspective*, Alexandria, Va.: Center for Naval Analyses, June 1987.

Defense Contract Management Agency, "About the Agency," web page, undated. As of April 2, 2016:
http://www.dcma.mil/about.cfm

Defense Intelligence Agency, "Mission Areas," web page, undated. As of April 2, 2016:
http://www.dia.mil/Careers/MissionAreas.aspx

Defense Logistics Agency, "DLA at a Glance," web page, undated. As of April 2, 2016:
http://www.dla.mil/AtaGlance.aspx

Defense Logistics Agency Instruction 3000.03, *Global Force Management*, January 4, 2013.

Defense Procurement and Acquisition Policy, "Areas of Responsibility," web page, last updated November 5, 2015. As of April 2, 2016:
http://www.acq.osd.mil/dpap/pacc/cc/areas_of_responsibility.html

Defence Science and Technology Group, homepage, undated. As of April 2, 2016:
http://www.dsto.defence.gov.au/

———, "About DST Group," web page, undated. As of April 2, 2016:
http://www.dsto.defence.gov.au/discover-dsto/about-dst-group

———, "Operations," web page, undated. As of April 2, 2016:
http://www.dsto.defence.gov.au/discover-dsto/our-role/operations

———, "Our Role," web page, undated. As of April 2, 2016:
http://www.dsto.defence.gov.au/discover-dsto/our-role

Defence Science and Technology Laboratory, "DSTL Support to Operations," RSS channel, last updated July 28, 2014. As of April 2, 2016:
http://operations1231.rssing.com/chan-7566085/all_p1.html

Defense Science Board, *Report of the Task Force on Human Resources Strategy*, Washington, D.C.: U.S. Department of Defense, February 2000.

———, *Final Report of the Task Force on Deployment of Members of the National Guard and Reserve in the Global War on Terrorism*, Washington, D.C.: U.S. Department of Defense, September 2007.

Defense Security Cooperation Agency, "Ministry of Defense Advisors," web page, undated. As of April 2, 2016:
http://www.dsca.mil/programs/ministry-defense-advisors

DeYoung, Karen, "State Department to Order Diplomats to Iraq," *Washington Post*, October 27, 2007a. As of April 2, 2016:
http://www.washingtonpost.com/wp-dyn/content/article/2007/10/26/AR2007102602417.html

———, "Envoys Resist Forced Iraq Duty," *Washington Post*, November 1, 2007b. As of April 2, 2016:
http://www.washingtonpost.com/wp-dyn/content/article/2007/10/31/AR2007103101626.html

———, "State Dept. Won't Order Diplomats to Iraq," *Washington Post*, November 16, 2007c. As of April 2, 2016:
http://www.washingtonpost.com/wp-dyn/content/article/2007/11/15/AR2007111501767.html

DoD—*See* U.S. Department of Defense.

DoDD—*See* U.S. Department of Defense Directive.

DoDI—*See* U.S. Department of Defense Instruction.

Drug Enforcement Administration, homepage, undated. As of April 2, 2016:
http://www.dea.gov/index.shtml

Dunigan, Molly, *Considerations for the Civilian Expeditionary Workforce: Preparing to Operate Amidst Private Security Contractors*, Santa Monica, Calif.: RAND Corporation, OP-359-OSD, 2012. As of April 2, 2016:
http://www.rand.org/pubs/occasional_papers/OP359.html

Eckholm, Erik, "U.S. Contractor Found Guilty of $3 Million Fraud in Iraq," *New York Times*, March 10, 2006. As of April 2, 2016:
http://www.nytimes.com/2006/03/10/international/middleeast/10contract.html

England, Gordon, Deputy Secretary of Defense, "Implementation of Section 324 of the National Defense Authorization Act for Fiscal Year 2008 (FY 2008 NDAA)—Guidelines and Procedures on In-Sourcing New and Contracted Out Functions," memorandum, April 4, 2008. As of April 2, 2016:
http://prhome.defense.gov/Portals/52/Documents/RFM/TFPRQ/Docs/OSD%20 In-sourcing%20Guidance%2004184-08.pdf

ENTRi, homepage, undated. As of November 4, 2015:
http://www.entriforccm.eu

European External Action Service, "About CSDP—Overview," web page, undated. As of April 2, 2016:
http://www.eeas.europa.eu/csdp/about-csdp/index_en.htm

———, "Civilian Crisis Management," web page, undated. As of April 2, 2016:
http://www.eeas.europa.eu/csdp/civilian-crisis-management/index_en.htm

———, "Security and Defence—CSDP," web page, undated. As of April 2, 2016:
http://www.eeas.europa.eu/csdp/

———, "Ongoing Missions and Operations," web page, October 2015. As of April 2, 2016:
http://www.eeas.europa.eu/csdp/missions-and-operations/index_en.htm

European Parliament, Directorate-General for External Policies, Policy Department, *CSDP Missions and Operations: Lessons Learned Processes*, Luxembourg: European Union Publications Office, April 2012. As of April 2, 2016:
http://www.tepsa.eu/download/CSDP%20Missions%20and%20Operations-%20 Lessons%20Learned%20Processes%20(DG-%20External%20Policies).pdf

European Union, "Civilian Headline Goal 2010," approved by the ministerial Civilian Capabilities Improvement Conference and noted by the General Affairs and External Relations Council, November 19, 2007, Document 14823/07. As of April 2, 2016:
http://www.consilium.europa.eu/uedocs/cmsUpload/Civilian_Headline_Goal_2010.pdf

Executive Office of the President, *National Security Strategy*, Washington, D.C.: White House, May 2010. As of April 2, 2016:
http://www.whitehouse.gov/sites/default/files/rss_viewer/national_security_strategy.pdf

Federal Emergency Management Agency, "Innovative Cost-Saving Partnership to Strengthen Disaster Response and Expand Opportunities for Young People," fact sheet, undated. As of April 2, 2016:
http://www.fema.gov/media-library-data/20130726-1826-25045-2512/nccc_fema_partnership_fact_sheet.pdf

———, "Community Emergency Response Teams," web page, last updated March 8, 2016. As of April 2, 2016:
http://www.fema.gov/community-emergency-response-teams

———, "About the Agency," web page, last updated April 2, 2016. As of April 2, 2016:
http://www.fema.gov/about-agency

GAO—*See* U.S. General Accounting Office (before July 7, 2004); U.S. Government Accountability Office (after July 7, 2004).

Gates, Robert, Secretary of Defense, "Utilization of the Total Force," memorandum, January 19, 2007.

———, "UCMJ Jurisdiction over DoD Civilian Employees, DoD Contractor Personnel, and Other Persons Serving with or Accompanying the Armed Forces Overseas During Declared War and in Contingency Operations," memorandum, March 10, 2008.

Gates, Susan M., and Albert A. Robbert, *Comparing the Costs of DoD Military and Civil Service Personnel*, Santa Monica, Calif.: RAND Corporation, MR-980-OSD, 1998. As of April 2, 2016:
http://www.rand.org/pubs/monograph_reports/MR980.html

Glanz, James, and Alissa J. Rubin, "From Errand to Fatal Shot to Hail of Fire to 17 Deaths," *New York Times*, October 3, 2007.

Global Affairs Canada, "About the Stabilization and Reconstruction Task Force," web page, last updated January 27, 2014. As of April 2, 2016:
http://www.international.gc.ca/START-GTSR/about-a_propos.aspx?lang=eng&_ga=1.206752979.1129131345.1415643543

Harrison, Todd, *Chaos and Uncertainty: The FY 14 Defense Budget and Beyond*, Washington, D.C.: Center for Strategic and Budgetary Assessments, October 24, 2013.

Hicks, Kathleen, Deputy Under Secretary of Defense for Strategy, Plans, and Forces, testimony before the Commission on Wartime Contracting in Iraq and Afghanistan at the hearing "Total Force Policy, the QDR, and Other Defense and Operational Planning: Why Does Planning for Contractors Continue to Lag?" Washington, D.C., July 12, 2010.

International Security Advisory Board, *Report on Security Capacity Building*, Washington, D.C.: U.S. Department of State, January 7, 2013, As of April 2, 2016:
http://www.state.gov/documents/organization/202920.pdf

"J1 Spreadsheets for CCMD Requirements for Deployable DoD Civilians," undated, provided to RAND by the Civilian Expeditionary Workforce Program Office in April 2014.

Joint Forces Staff College, "Introduction to Global Force Management," briefing on the Joint Operations Planning Process, March 3, 2012.

Kem, Jack, "The Ministry of Defense Advisor (MoDA) Program: MoDA as a Component of the Civilian Expeditionary Workforce," in *From Mission Creep to Mission Gallop: Changing the Dynamics in Afghanistan*, NATO Training Mission–Afghanistan, 2011.

Laird, Melvin O., Secretary of Defense, "Readiness of the Selected Reserve," memorandum, August 21, 1970.

Locher, James, *Victory on the Potomac: The Goldwater-Nichols Act Unifies the Pentagon*, Austin, Tex.: Texas A&M University Press, 2004.

Manuel, Kate M., and Jack Maskell, *Insourcing Functions Performed by Federal Contractors: An Overview of the Legal Issues*, Washington, D.C.: Congressional Research Service, November 15, 2011.

———, *Insourcing Functions Performed by Federal Contractors: Legal Issues*, Washington, D.C.: Congressional Research Service, February 22, 2013.

McGinn, Gail H., Deputy Under Secretary of Defense for Plans, "Department of Defense (DoD) Civilian Support to Global Expeditionary Requirements," memorandum, June 4, 2009.

"Memorandum of Agreement Between the Departments of State and Defense on the Protection and Evacuation of U.S. Citizens and Nationals and Designated Other Persons from Threatened Areas Overseas," July 1998. As of April 2, 2016:
http://prhome.defense.gov/Portals/52/Documents/PR%20Docs/DOS-DOD%20Memo%20of%20Agreement%20on%20Protection%20and%20Evacuation.pdf

National Security Presidential Directive 44, *Management of Interagency Efforts Concerning Reconstruction and Stabilization*, December 7, 2005.

Obama, Barack, President of the United States, "Government Contracting," memorandum, Washington, D.C.: White House, Office of the Press Secretary, March 4, 2009.

———, *Establishment of Afghanistan and Pakistan Strategic Partnership Office and Amendment to Executive Order 12163*, Executive Order, Washington, D.C., January 17, 2014. As of April 2, 2016:
http://www.whitehouse.gov/the-press-office/2014/01/17/executive-order-establishment-afghanistan-and-pakistan-strategic-partner

Office of the Assistant Secretary of Defense for Logistics and Materiel Readiness, "Future Combatant Command Requirements for Deployable DoD Civilians," memorandum, April 11, 2013.

Office of the Assistant Secretary of Defense for Readiness and Force Management, "Civilian Expeditionary Workforce Program Office Business Rules," memorandum, December 2, 2013.

Office of the Deputy Assistant Secretary of Defense for Civilian Personnel Policy/ Defense Civilian Personnel Advisory Service, "Civilian Expeditionary Workforce (CEW) Comprehensive Review: Summary," briefing slides, October 31– November 2, 2012.

———, "Deployable Civilian Demand Signal," briefing slides, April 23, 2014.

Office of Personnel Management, "Hiring Authorities: Competitive Hiring," web page, undated. As of April 2, 2016:
http://www.opm.gov/policy-data-oversight/hiring-authorities/competitive-hiring

Office of the Special Inspector General for Afghanistan Reconstruction and U.S. Department of State Office of Inspector General, *The U.S. Civilian Uplift in Afghanistan Has Cost Nearly $2 Billion, and State Should Continue to Strengthen Its Management and Oversight of the Funds Transferred to Other Agencies*, SIGAR Audit-11-17 and State OIG Aud/SI-11-45 Civilian Uplift, September 8, 2011.

Perito, Robert, *The U.S. Experience with Provincial Reconstruction Teams in Afghanistan: Lessons Identified*, Special Report 152, Washington, D.C.: United States Institute of Peace, October 2005.

———, *Provincial Reconstruction Teams in Iraq*, Special Report 185, Washington, D.C.: United States Institute of Peace, March 2007.

Professional Services Council, letter to the Secretary of Defense, May 3, 2010. As of April 2, 2016:
http://www.govexec.com/pdfs/050410rb1b.pdf

Public Law 109-163, National Defense Authorization Act for Fiscal Year 2006, January 6, 2006.

Public Law 109-364, National Defense Authorization Act for Fiscal Year 2007, October 17, 2006.

Public Law 111-84, National Defense Authorization Act for Fiscal Year 2010, October 28, 2009.

Public Law 111-383, National Defense Authorization Act for Fiscal Year 2011, January 7, 2011.

Public Law 112-81, National Defense Authorization Act for Fiscal Year 2012, December 31, 2011.

Riposo, Jessie, Irv Blickstein, Stephanie Young, Geoffrey McGovern, and Brian McInnis, *A Methodology for Implementing the Department of Defense's Current In-Sourcing Policy*, Santa Monica, Calif: RAND Corporation, TR-944-NAVY, 2011. As of April 2, 2016:
http://www.rand.org/pubs/technical_reports/TR944.html

Schorn, Daniel, "Billions Wasted in Iraq?" *60 Minutes*, February 12, 2006. As of April 2, 2016:
http://www.cbsnews.com/news/billions-wasted-in-iraq

Sellers, Cameron, *Provincial Reconstruction Teams: Improving Effectiveness,* thesis, Monterey, Calif.: Naval Postgraduate School, September 2007.

Sweeney, Patrick, *A Primer for: Guidance for Employment of the Force (GEF), Joint Strategic Capabilities Plan (JSCP), the Adaptive Planning and Execution (APEX) System, and Global Force Management (GFM)*, Newport, R.I.: U.S. Naval War College, Joint Military Operations Department, July 2013.

Tamburrino, Pat M., Jr., Deputy Assistant Secretary of Defense for Civilian Personnel Policy, written statement before the Committee on Armed Services, U.S. House of Representatives, at the hearing "Human Capital Management: A High Risk Area for the Department of Defense," July 14, 2011.

Thompson, Loren, "Pentagon Insourcing Binge Begins to Unravel," *Forbes*, March 7, 2011.

UK Ministry of Defence, "Preparing Civilian Staff for Operational Deployments," September 14, 2012. As of April 2, 2016:
https://www.gov.uk/government/news/preparing-civilian-staff-for-operational-deployments

U.S. Agency for International Development, "The Office of Crisis Surge Support Staff," web page, last updated May 12, 2014. As of April 2, 2016:
http://www.usaid.gov/who-we-are/organization/bureaus/bureau-democracy-conflict-and-humanitarian-assistance/office-2

———, "Office of Transition Initiatives: Bureau for Democracy, Conflict, and Humanitarian Assistance," web page, last updated July 10, 2014. As of April 2, 2016:
http://www.usaid.gov/who-we-are/organization/bureaus/bureau-democracy-conflict-and-humanitarian-assistance/office-1

———, "Where We Work," web page, last updated February 24, 2016. As of April 2, 2016:
http://www.usaid.gov/political-transition-initiatives/where-we-work

———, "Political Transition Initiatives," web page, last updated March 21, 2016. As of April 2, 2016:
http://www.usaid.gov/political-transition-initiatives

U.S. Air Force Personnel Center, CEW Program Manager, "Military-Civilian Comparison for Six-Month Deployment to OIF/OEF," provided to RAND on July 21, 2014.

U.S. Army Corps of Engineers, "About Us," web page, undated. As of November 11, 2015:
http://www.usace.army.mil/About.aspx

———, "Disaster Response Missions and Information," web page, undated. As of April 2, 2016:
http://www.usace.army.mil/Missions/EmergencyOperations/NationalResponseFramework.aspx

———, "Field Force Engineering," web page, undated. As of April 2, 2016:
http://www.usace.army.mil/Missions/MilitaryMissions/FieldForceEngineering.aspx

———, "Interagency and International Support," web page, undated. As of April 2, 2016:
http://www.usace.army.mil/Missions/MilitaryMissions/InteragencyInternationalSupport.aspx

———, "Offices and Units," web page, undated. As of April 2, 2016:
http://www.usace.army.mil/About/OfficesandUnits.aspx

U.S. Code, Title 5, Chapter 33, Examination, Selection, and Placement.

———, Chapter 55, Pay Administration.

———, Chapter 57, Travel, Transportation, and Subsistence.

U.S. Code, Title 10, Section 2333, Joint Policies on Requirements Definition, Contingency Program Management, and Contingency Contracting.

———, Subtitle A, Chapter 3, Section 129a, Armed Forces: General Policy for Total Force Management, 2011, p. 111.

U.S. Customs and Border Protection, "About CBP," web page, undated(a). As of April 2, 2016:
http://www.cbp.gov/about

———, "Along U.S. Borders," web page, undated(b). As of April 2, 2016:
http://www.cbp.gov/border-security/along-us-borders

———, "At Ports of Entry," web page, undated(c). As of April 2, 2016:
http://www.cbp.gov/border-security/ports-entry

———, "CBP Addresses Humanitarian Challenges of Unaccompanied Child Migrants," web page, undated(d). As of April 2, 2016:
http://www.cbp.gov/border-security/humanitarian-challenges

———, "From the Air and Sea," web page, undated(e). As of April 2, 2016:
http://www.cbp.gov/border-security/air-sea

———, "International Initiatives," web page, undated(f). As of April 2, 2016: http://www.cbp.gov/border-security/international-initiatives

U.S. Department of Defense, *Quadrennial Defense Review Report*, Washington, D.C., February 6, 2006.

———, "DoD News Briefing with Secretary Gates from the Pentagon," transcript, April 6, 2008.

———, *Quadrennial Defense Review Report*, Washington, D.C., February 2010.

———, *Sustaining U.S. Global Leadership: Priorities for 21st Century Defense*, Washington, D.C., January 5, 2012.

———, *Quadrennial Defense Review 2014*, Washington, D.C., March 4, 2014.

U.S. Department of Defense Directive 1100.4, *Guidance for Manpower Management*, February 12, 2005.

U.S. Department of Defense Directive 1200.7, *Screening the Ready Reserve*, Enclosure 2, November 18, 1999, certified current as of November 21, 2003.

U.S. Department of Defense Directive 1400.31, *DoD Civilian Work Force Contingency and Emergency Planning and Execution*, April 28, 1995.

U.S. Department of Defense Directive 1404.10, *DoD Civilian Expeditionary Workforce*, January 23, 2009.

U.S. Department of Defense Directive 3000.05, *Military Support for Stability, Security, Transition, and Reconstruction (SSTR) Operations,* November 28, 2005.

U.S. Department of Defense Directive 5100.3, *Support of the Headquarters of Combatant and Subordinate Joint Commands*, November 15, 1999, certified current as of March 2004.

U.S. Department of Defense Directive 5101.1, *DoD Executive Agent*, September 3, 2002, certified current as of November 21, 2003.

U.S. Department of Defense Instruction 1100.22, *Policy and Procedures for Determining Workforce Mix*, April 12, 2010.

U.S. Department of Defense Instruction 1400.25, *DoD Civilian Personnel Management System, Vol. 250: Civilian Strategic Human Capital Planning*, November 18, 2008.

———, *DoD Civilian Personnel Management System, Vol. 1230: Employment in Foreign Areas and Employee Return Rights*, July 26, 2012.

U.S. Department of Defense Instruction 1400.32, *DoD Civilian Work Force Contingency and Emergency Planning Guidelines and Procedures*, April 24, 1995.

U.S. Department of Defense Instruction 3000.05, *Stability Operations*, September 16, 2009.

U.S. Department of Defense Instruction 6490.03, *Deployment Health*, August 11, 2006.

U.S. Department of Homeland Security, *U.S. Department of Homeland Security Agency Financial Report: Fiscal Year 2013*, Washington, D.C., December 2013. As of April 2, 2016:
http://www.dhs.gov/sites/default/files/publications/dhs_agency_financial_report_fy2013_vol1.pdf

———, *Budget-in-Brief, Fiscal Year 2015*, Washington, D.C., 2014. As of April 2, 2016:
http://www.dhs.gov/sites/default/files/publications/FY15BIB.pdf

U.S. Department of State, "Bureau of Conflict and Stabilization Operations," web page, undated. As of April 2, 2016:
http://www.state.gov/j/cso

———, "Bureau of Diplomatic Security," web page, undated. As of April 2, 2016:
http://www.state.gov/m/ds

———, "Bureau of South and Central Asian Affairs," web page, undated. As of April 2, 2016:
http://www.state.gov/p/sca

———, "Diplomatic Security Service Field Offices," web page, undated. As of April 2, 2016:
http://www.state.gov/m/ds/about/c8849.htm

———, *Foreign Affairs Manual*, 2 F.A.H.-2 H-112, "Foreign Assistance Standardized Program Structure and Definitions," April 8, 2010. As of April 2, 2016:
http://www.state.gov/documents/organization/141836.pdf

———, *Foreign Affairs Manual*, 2 FAH-2 H-100, "Post Management Organization and Chief of Mission Authority and Overseas Staffing," July 18, 2014. As of April 2, 2016:
http://www.state.gov/documents/organization/89604.pdf

U.S. Department of State, Bureau of Conflict and Stabilization Operations, *CSO at Two Years: Engaging Around the World*, Washington, D.C., March 13, 2014. As of April 2, 2016:
http://m.state.gov/md223397.htm

U.S. General Accounting Office, *High-Risk Series: An Update*, Washington, D.C., GAO-01-263, January 2001.

———, *DoD Personnel: DoD Actions Needed to Strengthen Civilian Human Capital Strategic Planning and Integration with Military Personnel and Sourcing Decisions*, Washington, D.C., GAO-03-475, 2003.

U.S. Government Accountability Office, *The Department of Defense's Civilian Human Capital Strategic Plan Does Not Meet Most Statutory Requirements*, Washington, D.C., GAO-08-439R, February 6, 2008.

———, *Defense Infrastructure: Army and Marine Corps Grow the Force Construction Projects Generally Support the Initiative*, Washington, D.C., GAO-08-375, March 2008.

———, *Provincial Reconstruction Teams in Afghanistan and Iraq*, Washington, D.C., GAO-09-86R, October 1, 2008.

———, *Human Capital: Opportunities Exist to Build on Recent Progress to Strengthen DoD's Civilian Human Capital Strategic Plan*, Washington, D.C., GAO-09-235, February 2009.

———, *Human Capital: Further Actions Needed to Enhance DoD's Civilian Strategic Workforce Plan,* Washington, D.C., GAO-10-814R, September 27, 2010.

———, *Afghanistan: Improvements Needed to Strengthen Management of U.S. Civilian Presence*, Washington, D.C., GAO-12-285, February 27, 2012.

———, *High-Risk Series: An Update*, Washington, D.C., GAO-13-283, February 2013.

———, *Human Capital: Additional Steps Needed to Help Determine the Right Size and Composition of DoD's Total Workforce*, Washington, D.C., GAO-13-470, May 2013.

———, *Human Capital: Opportunities Exist to Further Improve DoD's Methodology for Estimating the Costs of Its Workforces,* Washington, D.C., GAO-13-792, September 25, 2013.

U.S. House of Representatives, Committee on Armed Services, Subcommittee on Oversight and Investigations, "Deploying Federal Civilians to the Battlefield: Incentives, Benefits, and Medical Care," hearing, April 2008.

U.S. Joint Chiefs of Staff, *The National Military Strategy of the United States of America: Redefining America's Military Leadership*, Washington, D.C., February 2011a.

———, *Joint Operational Planning*, Joint Publication 5-0, August 11, 2011b.

U.S. Joint Forces Command, Joint Warfighting Center, "Provincial Reconstruction Teams," Pre-Doctrinal Research White Paper No. 07-01, November 21, 2007.

U.S. Navy, *Navy's Total Force Vision for the 21st Century*, January 2010. As of April 2, 2016:
http://www.navy.mil/features/NTF%20Vision%20(Final)(11%20Jan%2010%20 1210hrs).pdf

U.S. Senate, Committee on Armed Services, "Department of Defense Authorization of Appropriations for Fiscal Year 2015 and the Future Years Defense Program: The National Commission on the Structure of the Air Force," hearing, Washington, D.C., April 29, 2014.

Walker, David M., Comptroller General of the United States, *Defense Management: DoD Needs to Reexamine Its Extensive Reliance on Contractors and Continue to Improve Management and Oversight*, testimony before the Subcommittee on Readiness, Committee on Armed Services, U.S. House of Representatives, Washington, D.C., GAO-08-572T, March 2008.

Weigelt, Matthew, "Army Vows to Cut 7,000 Contractor Jobs This Year," *Washington Technology*, February 23, 2010. As of April 2, 2016: http://washingtontechnology.com/articles/2010/02/23/army-insourcing-core-governmental-functions.aspx